Early Stationary Steam Engines in America:
A Study in the Migration of a Technology

Carroll W. Pursell, Jr.

Early Stationary Steam Engines in America

A STUDY IN THE MIGRATION OF A TECHNOLOGY

SMITHSONIAN INSTITUTION PRESS
CITY OF WASHINGTON
1969

Library of Congress catalog card number 73-600455

Distribution in the United States and Canada by Random House, Inc.
Standard Book Number 87474-094-0

Distribution in the United Kingdom and Europe by
David & Charles (Publishers) Ltd., South Devon House,
Newton Abbot, Devon, England.

Contents

Introduction

This book is an attempt to trace the history of the introduction of stationary steam engines into the United States. It is a history which has been largely neglected for two reasons. First, the drama and glamour of steamboats and railroads have eclipsed the less spectacular subject of stationary engines. This was true even during the period when all three were being absorbed into, and transformed by, their new American environment, and that contemporary neglect has influenced later historians.

The second reason for this subject's neglect is that the history of the Industrial Revolution in the United States has been written mainly in terms of the economic development of New England. Generalizations based upon the study of this region, which was unique in so many ways, have been accepted as applicable to the rest of the country. There have been recurrent statements to the effect that steam power in industry did not become important until after the Civil War. These may be traced to the fact that such large New England textile centers as Lawrence and Lowell, Massachusetts, continued to depend primarily upon waterpower during the first half of the nineteenth century and long thereafter. Such generalizations are not true of the rest of the country, or even of the rest of New England.

This neglect is unjustified. It has long been acknowledged that the advent of steam power and an increasingly mechanized textile industry was the great motive force behind the Industrial Revolution, which began in England late in the eighteenth century, spread to this country and continental Europe, and changed the face of the land. Through the seventeenth century, England had been a commercial, agricultural, and handicraft-manufacturing nation. The revolution began with the application of waterpower to textile machinery and was made permanent with the substitution of steam engines for water wheels. The British Isles, exploiting their technological lead, fashioned industry into a powerful weapon of imperialism, and with it won a century of world leadership.

It would be unrealistic to look in America for the identical effects which the introduction of steam power brought about in England. But it would be equally naive to assume that steam had neither cause nor effect on this side of the Atlantic. This book does not pretend to be a complete history of the American economy before the Civil War. Neither is it primarily a history of the stationary steam engine, a story which is essentially English and has been told many times from English sources. It is rather my purpose to show when and by whom steam engines were brought to America; the extent and nature of their application here; and, so far as possible, the changes which they imposed upon the existing patterns of American industry.

Because this is basically a narrative history, I have relied mainly upon the traditional source of such history: the written document. The appended bibliography will indicate those sources which I found most useful in making this study, but a special word should be said about statistics. Since the mid-nineteenth century, the federal government has become a vast source and repository of statistics concerning every facet of our country's life. For the pre-Civil War period, however, the student of American history has no statistics, but only scattered figures. Certain "manufactured" statistics have been produced, but they seldom bear any real relation to fact. Where figures have been available, I have presented them for what they are worth. At the same time, I have eschewed conventional comparisons rather than base them on statistics made for the occasion.

My general rule concerning definitions has been to use words in their commonsense, historical meanings. In most cases I expect this practice to cause a minimum of confusion. Modern engineers, for example, deal with steam of supercritical pressures, that is, over 3,226 psi (pounds per square inch)—an amount that could only be dreamed of in the early nineteenth century. The first "atmospheric" steam engines used pressures of less than one atmosphere, that is, less than about 15 psi. Such steam was introduced into the cylinder merely to be condensed, thus creating a vacuum so that the air pressure above the piston might press it downward. Early in the nineteenth century when Oliver Evans and others developed engines which dispensed with condensers and used steam directly to push a piston, steam of greater pressure than the surrounding air was necessarily used. Such engines were called high-pressure engines (over 15 psi) in contrast to the old, condensing, low-pressure (less than 15 psi) engines. In practice, the "high" pressure of the Evans engine was not very high—usually less than 200 psi—and the "low" pressure of the Watt or Newcomen engine, at least in the United States, was seldom less than 15 psi. The pressure chosen by an engineer in any given situation was determined by such considerations as the strength of his boiler and the relative demands of work level and efficiency, rather than by the nominal designation of his engine. Since actual operating pressures were always difficult to determine, while general construction was usually obvious, in the early nineteenth century engines with condensers were referred to as low-pressure engines and all others as high-pressure engines. In this book the terms high and low are used in that sense.

Abbreviations used in the footnotes are:

LC	Library of Congress
MHS	Maryland Historical Society
MS. Coll.	Manuscript Collection
NYHS	New York Historical Society

I am indebted to the several archives and libraries which have generously allowed me to consult and quote materials in their possession, and particularly the Maryland Historical Society for use of their Benjamin Henry Latrobe manuscript collection. Professor A. Hunter Dupree encouraged this work from its inception. Professors Charles G. Sellers, Jr., Paul F. Keim, James L. Penick, and Bruce Sinclair kindly read and criticized the manuscript, for which I am grateful. Special thanks are due Professor Eugene S. Ferguson, whose careful suggestions improved the text at many points. A timely grant from the University of California Committee on Research has aided my work greatly.

Carroll W. Pursell, Jr.
University of California
Santa Barbara

Chapter 1

The Colonial Experience

The mystery was, how the machine in the well was first actuated and kept in motion. This (as I at length discovered, with surprise) was done by the power of boiling water.

ISAAC BANGS (1776)[1]

By the mid-eighteenth century, Great Britain was just beginning to feel that quickening of social and economic conditions which would so astonish observers during the next century and earn for the period the name Industrial Revolution. Most Englishmen still lived on the land, engaged in rural occupations in a rustic setting. Although the main part of this population specialized in farming, a growing number were involved in industrial activities, for it was in the country rather than in the cities of the realm that manufacturing was pursued.

During the previous two centuries the manufactories of England had moved from the cities to the countryside. In part this was to escape the restrictions which the various guilds and municipalities imposed upon work carried on within corporate limits and to a degree it was dictated by the technical requirements of locating near materials and having access to the power provided by wind and water. Few towns had sufficient waterpower within their limits to support large industries, and the fuel requirements (both for coal and wood) of such establishments as ironworks were satisfied more easily away from the cities. In short, the widespread geographical distribution of natural resources led to extensively distributed industrial activity.

Along the banks of streams, small mills had been established to take advantage of the fall of water. Many of these were flour (corn) mills but as an iron industry burgeoned in the eighteenth century, quite a few were converted to other purposes. This was true particularly in south Staffordshire, northwest Worcestershire, and other areas which were becoming centers of iron production. Furnaces to smelt iron from ore, forges to hammer cast into wrought iron, slitting mills to roll and cut iron into bands and rods, and cutlery and tool-grinding mills stood alongside the millstreams of the countryside. Fulling mills, designed to shrink and mat newly woven cloth, and other types of manufactories were also common. The power available for these mills was seldom more than ten horsepower and probably averaged closer to five at each millsite. Such limited power resulted from comparatively inefficient water wheels, heavy and crude gearing, and the low gradient of most streams in England. Though limited, this power was ample for most industrial enterprises of that day.

Americans traveling abroad often carefully noted foreign industrial activity and observed how it differed from American practices. In 1785 the Connecticut merchant and sometime diplomat Silas Deane reported that "there are at present no Saw Mills in England, but all the sawing is done by hand, by which many thousands of Men are employed."[2] An even more startling sight for an American was the large number of windmills employed. In 1793 Robert Leslie, a Philadelphia watchmaker, wrote back from London that

> the Milling business is carried on here, in a very different manner from what it is in America, that is, all in the small way, and almost entirely by wind mills, each of which does but very little business, but are able to do all the work of the Country in consequence of their great numbers, which exceeds any thing you can amagin. there is I believe hardly a piece of high ground in England, but is covered with them, and all along the rivers they are so thick that you may stand in many places and count mor than thirty . . . . I have not yet seen nor heard of a water mill, going by a fall of water, as falls are very scarce, and what there are, is used for forges, furnisces, plating, roling, grinding cuttlery, and many other kind of manufactorys. . . .[3]

So small and numerous were the mills of the kingdom that there were some five hundred in and around London in the mid-1780s, merely to supply that city with flour.[4] Despite differences of practice, the general picture was clearly one of an extensive industrial activity, based upon power-driven machinery.

Another sector of England's growing industrial economy that was tied to the rural area was coal mining. By 1700 the coal mining industry long since had passed the stage of exploiting easily accessible outcroppings or open and comparatively shallow pits. Already shafts had been sunk two hundred feet in depth, and water, the great enemy of successful mining, was a rapidly growing problem as spring after spring was ruptured underground. At the bottom of each mine were sumps of filthy water which somehow had to be removed. One of the oldest and best removal devices was the adit, a small tunnel which carried water from a hillside mine. The adit could drain only those areas where the source was higher than the outlet, but the method was used where feasible. Adits dating from the eighteenth century are still operative in Cornwall, and some extend a distance of fifty miles.

Where an adit was not possible, hand pumps were sometimes resorted to as was the rag-and-chain, a device made up of knots of rag attached to an endless chain which carried water up through a tube. The principal limiting factor in the use of these devices was a lack of sufficient power. Men, draft animals, windmills, and waterwheels all had to be used. Because animal power was relied upon so often, horses and donkeys that were needed to help raise water were kept on large farms maintained by mine operators as part of their work. Eventually, if sufficient power was not available to pump out water, the mines were simply forced to shut down operations, sometimes with large quantities of rich minerals trapped beneath the rising waters. While wind and water were still adequate to provide for the power needs of English manufacturers, the miners found themselves facing something of a crisis.

The solution for the growing power needs was finally achieved with the invention of the steam engine by Thomas Newcomen, an ironmonger of Dartmouth in Devonshire. Details of the origins of his epic invention are unknown. He was a man of wide practical experience coupled with a quick and fertile intelligence. Certainly he was familiar with the nearby tin mines and their water problems, though it appears unlikely that he was aware of the studies of air pressure and vacuums which had played such an important part in the scientific revolution of the previous century. We know little more than that by 1712 Newcomen had worked out the combination and proportions of a steam engine and had one in operation in Staffordshire.

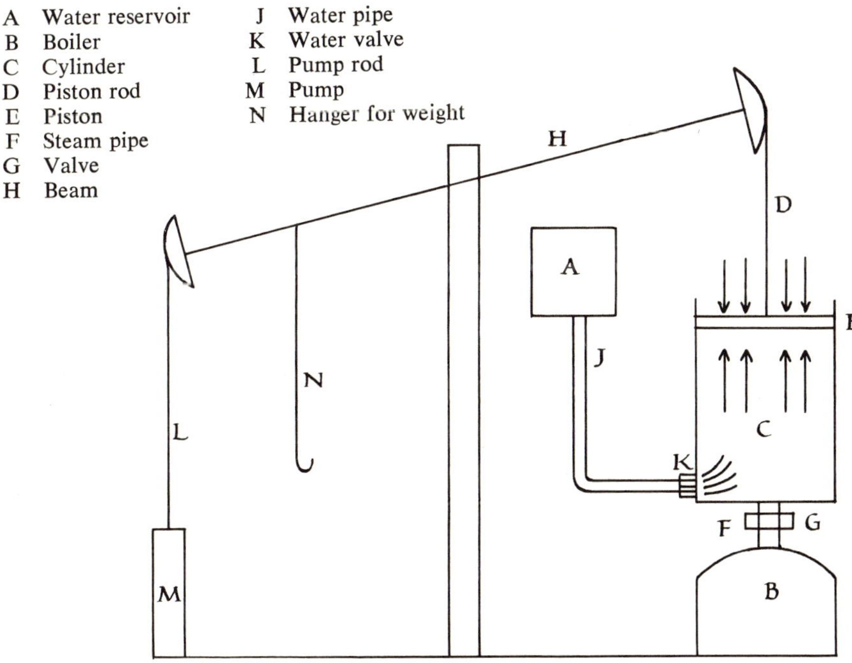

Figure 1 Newcomen engine, about 1750

Figure 1 represents the main parts of a newcomen engine as it was built in the mid-eighteenth century. When the engine is at rest, the piston (E) is at the top of the steam cylinder (C), since the combined weight of the piston and piston rod (D) is less than that of the pump (M) and pump rod (L). To start the cycle of the engine, the valve (G) is opened, allowing steam to rise from the boiler (B), through the pipe (F), into the cylinder. When the pressure of the steam within the cylinder has driven out the air formerly there, the valve (G) is closed. The other valve (K) is then opened, allowing cold water to flow from the reservoir (A), through the pipe (J), and to spray into the cylinder. The spray of cold water condenses the steam in the cylinder, producing a partial vacuum. The pressure of the atmosphere above the cylinder (about 15 psi) is now added to the weight of the piston and rod, making this end of the beam heavier and therefore driving it down, raising the pump end and with it the water within the pump.

The cycle is completed by again opening the valve (G) and relieving the vacuum in the cylinder beneath the piston. As the pressure approaches that of the atmosphere, the pump end of the beam, again relatively heavier, descends, raising the piston to its original position near the top of the cylinder.

This new engine was well suited to solve the water problems of eighteenth-century English mines. It provided power in necessary amounts, its reciprocating motion was ideal for the operation of pumps, and its great bulk and comparative inefficiency (in modern terms) did not detract from its usefulness in areas desperate for stationary power and well endowed with fuel.

It was not long after Newcomen's success that interested travelers were spreading word of the new invention to the continent of Europe. During the 1720s a number of engines were erected, with or without the aid of Englishmen, in several countries: the first at Königsberg in the mining district of Hungary by 1725, followed shortly by others at Vienna, and Cassel in Germany. An engine was erected in Spain to pump water for the city of Toledo; in 1726 the first in France was set up at Passy to pump water for Paris from the river Seine, and the following year the first engine in Sweden was established to pump water from a mine. By 1730 in all of western Europe, wherever there was a need for large quantities of water to be pumped from mines or to cities, the Newcomen engine was a possible though not always successful contender for the job of supplying power.

The existence of the Newcomen engine, and some idea of its workings, also became known in America. The young John Adams dutifully recorded in his diary in 1760 that he was struggling to understand the English "fire engines" and noted with his usual thoroughness that though formerly constructed of copper, lately they had been made of iron.[5] The interest elicited by the engine in the New World was limited largely to those who had a philosophical turn of mind and the leisure and opportunity to dabble in science. In later life Thomas Jefferson recalled that the steam engine, as first constructed by Newcomen, had been a subject of his early studies,[6] probably at the College of William and Mary. One of his instructors there was William Small, a professor of natural philosophy who later played an important part in bringing about the fruitful partnership between Matthew Boulton and James Watt. Benjamin Franklin's perpetual and ubiquitous curiosity led him also to make the acquaintance of Boulton in 1758 and they corresponded on the usefulness of steam power.

Any discussions of steam engines in America during colonial times were bound to be largely academic. The Newcomen engine, ponderous as it was, could find usefulness only in pumping water. Urban waterworks hardly were thought of in America yet, and cities were not large enough to make the need for them felt. Iron mines were being worked all along the American coast from Massachusetts to Virginia, but they were so new and the ore so plentiful that very few of them had reached any great depth.

In most places iron ore could be scraped from the bottom of bogs with oyster rakes or could be dug from open pits. While on an inspection trip of American mines in 1783, the Swedish engineer Samuel Gustaf Hermelin found that where water had to be removed from mines, simple expedients often were sufficient. In one Pennsylvania mine he reported a hand pump in use, and at a New Jersey mine he found that a small adit could carry away the water. In New Jersey, ore was so plentiful near the surface that mines were not worked if water accumulation made it difficult to proceed.[7] Under such conditions, there was little likelihood that steam engines could pay their way.

The situation was more critical with regard to scarcer metals. The first steam engine in America was finally erected in the Schuyler copper mine near Passaic, New Jersey. Rumors persist that two or three engines were used in New England before the Revolution, but in the absence of any specific information, they are better discounted.[8] In 1748 the copper mine of Colonel John Schuyler, located on the Passaic River just above Newark, became filled with enough water to threaten its further usefulness. When Benjamin Franklin visited this mine in the fall of 1748, he reported that it "yields good copper, and has turned out vast wealth to the owners." It was, however, "not then at work: the water is grown too hard for them." Colonel Schuyler, he continued, had ordered a "fire engine" from England (at the reported cost of £ 1,000) and hoped to have it in operation by the following summer.[9]

Schuyler, having heard of Newcomen's device, had requested his London agent to contract for an engine and to secure a workman to accompany it to America to oversee its erection and operation. The contract had been made with the Hornblowers, one of England's leading families of engineers. Joseph Hornblower, the head of the family, had been an early associate of Newcomen and in 1725 had moved with his family to Cornwall to erect engines in the mines. There two of his sons, Jonathan and Josiah, learned the engine-building business. Jonathan became a leading English engineer; his younger brother Josiah accompanied the Schuyler engine to America. In September 1753 he arrived in New Jersey, accompanied by the engine parts, many in duplicate and some even in triplicate, and a few mechanics.[10]

Apparently Hornblower's intention was to return to England when the Schuyler engine was set up and operating properly. This was not accomplished, however, until March 1755, a year and a half after he had arrived—and nearly seven years after the engine had been ordered. At this time, instead of returning to England, Hornblower accepted the offer of Colonel Schuyler to remain and superintend the operations of the mine and engine.

Operations at the mine were uneventful until fire destroyed the engine about 1760.[11] Some of the parts needed to repair it were ordered from England and others were cast at a foundry near the mine. Hornblower and a partner leased the mine from Schuyler in July 1761 and were operating it when the engine burned again in 1768. Following this disaster, the engine was not rebuilt until after the Revolution. According to the traveler Johaan Schoepf, who visited the mine early in the 1790s, Hornblower had continued to operate by reclaiming copper from the waste of ore already mined and reduced over the years. Until the war halted operations, he was reclaiming seven to fifteen tons of pure metal annually by this method: "proof," wrote Schoepf, "of how carelessly the ore had been worked."[12] Hornblower meanwhile married into a prosperous local family, settled into a new home, and became financially interested in a nearby store and ferry.

The oddity of the large Newcomen engine made it a tourist attraction for the ubiquitous travelers, and with eighteenth-century seriousness of purpose each recorded the facts as he received them. In 1776 Isaac Bangs, a soldier from Massachusetts, visited the abandoned mine and was told that the engine had "cost about £ 3,000 sterling, and would cast out of the earth 80 hogsheads of water per minute, being actuated by fire, from which it had its only motion."[13] It was not until 1793 that another attempt was made to operate the mine. A "New Jersey Copper Mine Association" was formed by Philip A. Schuyler, Jacob Mark, and Nicholas I. Roosevelt. This company hired Hornblower to rebuild the engine, to erect a stamping mill (to crush the ore for smelting), a foundry, and a machine shop, and to superintend the enterprise. After Hornblower retired from active service the next year, the works were continued under Roosevelt's domination and were named Soho, after the works of Boulton and Watt in England.

Visiting the mine shortly after Hornblower severed his connections with it, the Duke de la Rochefoucault Liancourt reported that "the steam-engine for pumping off the water is very ill contrived, and the defect in its construction extends its influence to the working of the mine."[14] In 1804 Benjamin Henry Latrobe, who had a personal interest in the Soho Works, admitted that the last Hornblower repair had not been the best, but claimed that the engine, "notwithstanding its imperfect construction and faulty boring of its cylinder, effectively drained the mine."[15] Operations continued at the old Schuyler mine for many years after the death of Hornblower in 1809, and in 1847 the vein was being drained by a beam engine of 40 horsepower.[16]

The next steam engines to appear in America were even less successful, being entirely of colonial construction. These were built by the Irish immigrant Christopher Colles, later better known for his maps and travel guide. He had come to America in 1765 and by 1771 was in Philadelphia engaged in the trade of millwright. Colles began the construction of a small steam engine in 1773 to raise water for a Philadelphia distillery. He described it as "very little differing from the common construction." Before he had it operating to his complete satisfaction, however, he ran short of money. It was his intention to solicit funds from the public to complete the work, and he turned to Benjamin Franklin's American Philosophical Society for a testimonial to the competence of the machine and himself. "As all persons are not competent judges of the nature & utility of such contrivances," he wrote, "I request the favour of the society, that such gentlemen of their body as they shall think convenient may look at the machine, & report their opinion thereof, which will enable the subscription to go on with the greater facility."[17]

The Society appointed the astronomer David Rittenhouse and two other men to view the engine and to report back. They returned with the information that

> the Committee went to view the Steam-Engine erected by Christopher Colles in this city, which they saw perform several strokes (tho' some of the materials not being sufficiently large and strong, owing to his attempting the execution at a very low expense), it did not continue in motion long: but that a Steam-Engine may be brought to answer the purpose of raising water, much cheaper than by men or horses, is a fact well known, and we are of opinion that the undertaker is well acquainted with the principles of this particular branch of mechanics, & very capable of carrying it into execution, and therefore worthy of publick encouragement.[18]

This attempt to build an engine apparently came to nothing, probably through a continued lack of financial support. The picture of a committee of savants endorsing a public subscription to finance a private technological venture served as a fair measure of the ambiguity which still surrounded the steam engine in the colonies. It was at the same time a philosophical curiosity, a public work, and a device of potential economic advantage.

The year after his failure in Philadelphia, Colles moved to New York where he attempted to interest the city fathers in the ability of steam to raise sufficient water for the proposed city waterworks. The population of the city was only 22,000 in 1774, but already the exclusive reliance upon a multitude of private wells was proving inadequate. In July 1774 the Corporation of the city agreed to his terms, and Colles received his second chance to build a successful engine.[19] The cylinder for this engine was cast in New York at the air furnace of Peter T. Curtenius. The casting was thought at the time to be "the first performance of the kind ever attempted in America, and allowed by judges to be extremely well executed."[20]

The Newcomen-type engine was in operation by April 1776, when it was visited by Bangs, the young graduate of Harvard College then serving with the American forces in New York. After a close examination, he confessed that

> the mystery was, how the machine in the well was first actuated and kept in motion. This (as I at length discovered, with surprise,) was done by the power of boiling water. I found that, by means of a large copper boiler, the steam or vapor of the water is conveyed from thence into a strong copper tube of 18 inches and 10 feet long, which stands in a perpendicular position [i.e., the cylinder]. The lower

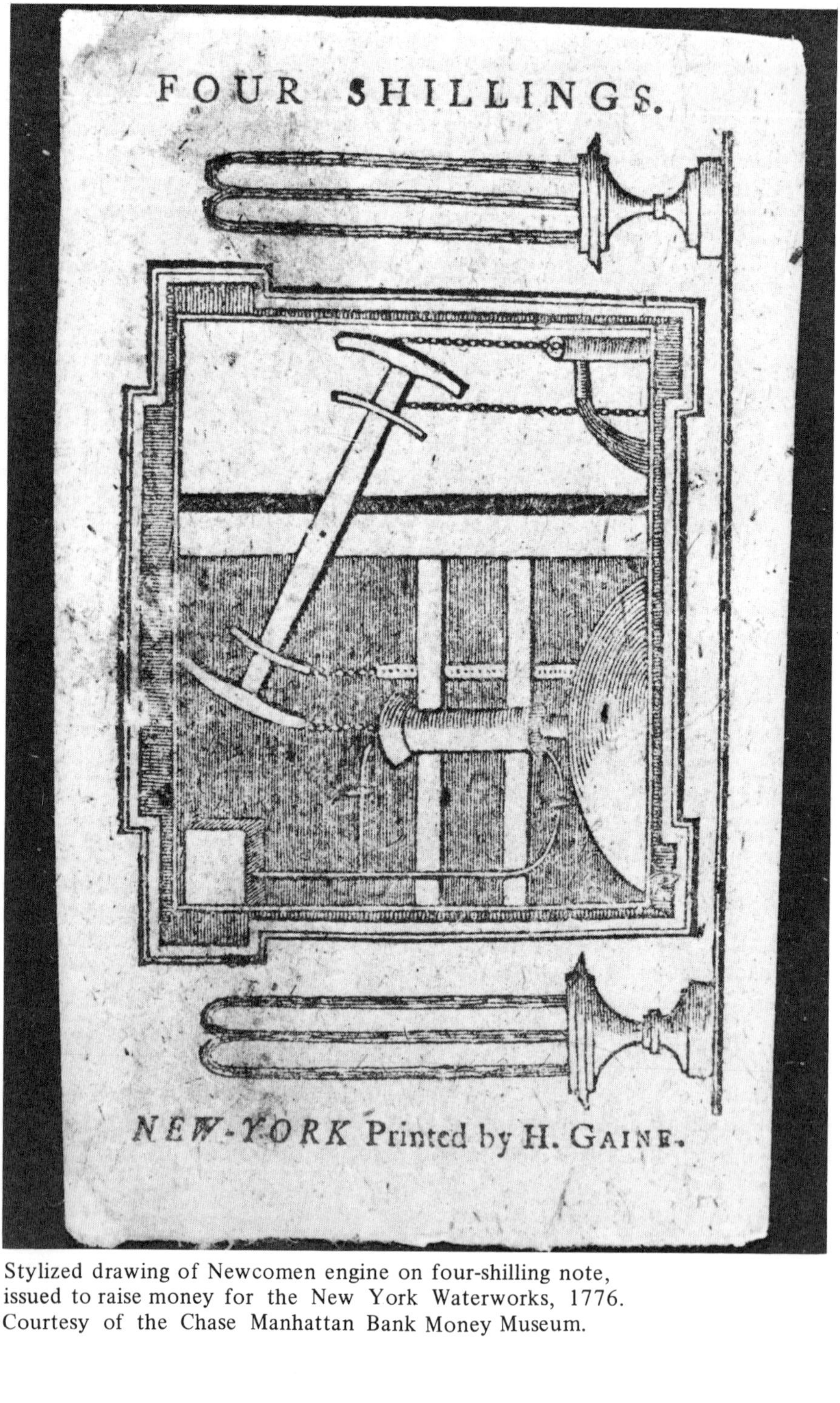

Stylized drawing of Newcomen engine on four-shilling note, issued to raise money for the New York Waterworks, 1776. Courtesy of the Chase Manhattan Bank Money Museum.

part or end of this tube is tight, but the upper end has in it a moveable stopper [the piston] which may move upwards or downwards with as much ease as possible, and at the same time keep the air without from entering into the tube. In order to keep it tight, another part of the works constantly supplies the top of the tube, above the stopper, with a small jet of water. The stream of the hot water (as I take it) entering the tube, rarifies the air therein to a great degree, when the stopper is let loose and flies upward with rapidity to the upper end of the tube, and immediately is thrown back by the pressure of the air from without. When it gets to the bottom, it is again driven upwards by the same cause as before, and repelled downward in like manner by the air, causing a constant motion. To this stopper a stout lever [beam] is fastened in the middle upon the axis. This lever is moved up and down by the stopper in the tube, and thus works the engine as well . . . .[21]

The British occupation of New York disrupted work on the water system, and, although he lived until 1821, Colles apparently had no further direct contact with the construction of steam engines.

The same Peter T. Curtenius who cast the cylinder for Colles's waterworks engine also played a part in the next American engine to be built, that of Joseph Brown at the Hope Furnace in Rhode Island. This iron furnace had been built by Samuel Waldo about 1735 near the town of Cranston. By 1767 the Browns, a merchant family of Providence, were interested in the furnace and in that year John Brown traveled through Pennsylvania and New Jersey looking for an iron founder willing to move to Rhode Island. It is possible that on this trip he may have seen the Hornblower engine at the Schuyler mine.[22]

In 1776 the Browns took an order for cannon from the government of Rhode Island, and Nicholas Brown asked his friend Curtenius to send workmen familiar with casting cannon to the Hope Furnace. Since this was only a year after Curtenius had cast the Colles cylinder, the men he sent were no doubt familiar with the problems involved. A French officer stationed in Rhode Island also instructed workmen at the furnace in the art of casting cannon.[23] The techniques and facilities used for casting and boring cannon were quite similar to those required for making a steam cylinder, and trained workmen were thus available in 1780 when Joseph Brown built the first steam engine to be seen in New England.

The best description of this engine is that left by the Reverend Manasseh Cutler after he visited the Brown enterprise in 1787. This "singular" machine, which Cutler thought to be the "Only one in America," was clearly of the Newcomen type and had a solid oak beam four feet square and twenty feet long. It was claimed to be able to raise seven hogsheads of water per minute from the iron mine, using only two cords of wood in twenty-four hours. Cutler noted that "the invention was not new," but reported that Joseph Brown had "made many valuable improvements, in simplifying and making the working of it more convenient, above what has yet been done in Europe."[24] Unfortunately, he did not specify what these improvements were. In 1808 the engine was still in operation, and one visitor observed, "a very useful monument" to the "mechanical genius" of its builder.[25]

These three engines–in New Jersey, Rhode Island, and New York–were the only steam engines erected in the country before the end of the Revolutionary period and were probably the only Newcomen engines ever set up on this side of the Atlantic. The reason had nothing to do with colonial ignorance or official discouragement by the British crown. There was simply no real demand for such engines, whose usefulness even in England was limited to the pumping of water. The American colonies provided neither the deep mines nor the large urban waterworks which would have made steam engines economically attractive. The Brown and Colles engines were constructed to provide for rather marginal needs; only the Schuyler engine provided an indispensable solution to a serious problem.

To erect a steam engine before the 1770s required a comparatively rare combination of economic need, entrepreneurial daring, considerable capital, and access to sufficient mechanical skill and fabricating facilities. It has been estimated that by 1775 in Great Britain only 130 steam engines were in operation, for all purposes. In America the problems of pumping water were not sufficient to encourage millwrights to experiment with the new technology of steam. When someone did, he became a "genius" and his machine a "marvel." The further introduction of the steam engine would have to wait until it held some promise of solving the more basic problems of the American economy.

Notes

[1] [Isaac Bangs], "New York in 1776," *Historical Magazine,* volume 4, 2d series (December, 1868), pages 305-306.

[2] Silas Deane to Simeon Deane, 20 May 1785, in *Correspondence between Silas Deane and His Brothers and Their Business and Political Associates, 1771-1795. Collections of the Connecticut Historical Society,* volume 23 (Hartford, 1930), page 212.

[3] Robert Leslie to O. Evans, 26 September 1793, Oliver Evans papers, Franklin Institute, Philadelphia, Pennsylvania.

[4] O. A. Westworth, "The Albion Steam Flour Mill," *Economic History* (supplement to *Economic Journal*), volume 2 (January 1932), page 383.

[5] L. H. Butterfield, editor, *Diary and Autobiography of John Adams* (Cambridge, 1961), volume 1, page 134, entry for 14 June 1760.

[6] Thomas Jefferson to George Fleming, 29 December 1815, in H. A. Washington, editor, *The Writings of Thomas Jefferson* (Washington, 1853-54), volume 6, page 504.

[7] Samuel Gustaf Hermelin, *Report About the Mines in the United States of America, 1783,* translated by Amandus Johnson (Philadelphia, 1931), pages 34, 22, 24.

[8] See for example, Thompson Westcott, *The Life of John Fitch* (Philadelphia, 1857), page 153n.

[9] Benjamin Franklin to Jared Eliot, 13 February 1749, quoted in *American Journal of Science,* volume 4 (1822), page 365.

[10] Rolt L.T.C., *Thomas Newcomen:* The Prehistory of the Steam Engine (London, 1963), pages 64, 122-124; "The History of the Steam Engine in America," *Journal of the Franklin Institute,* volume 102 (October 1876), pages 254-255; L. F. Loree, *First*

Steam ("Fire") Engine in America (no place, no date).

[11](Rind) *Virginia Gazette,* 18 August 1768.

[12]Johann David Schoepf, *Travels in the Confederation,* edited by Alfred J. Morrison (Philadelphia, 1911), volume 1, pages 33-34.

[13][Bangs], "1776," *Historical Magazine,* volume 4 (1868), page 306n.

[14]Duke de la Rochefoucault Liancourt, *Travels through the United States of North America . . . in the Years 1795, 1796, and 1797* (London, 1799), volume 2, page 244.

[15]Benjamin Henry Latrobe, "First Report of Benjamin Henry Latrobe, to the American Philosophical Society. . . ," *Transactions of the American Philosophical Society,* volume 6, part 1 (Philadelphia, 1804), page 90.

[16]William G. Lathrop, *The Brass Industry in Connecticut: A Study of the Origin and the Development of the Brass Industry in the Naugatuck Valley* (Shelton, Connecticut, 1909), page 43n; *Hunt's Merchants' Magazine,* volume 18 (January 1848), page 110.

[17]Christopher Colles to American Philosophical Society, 20 August 1773, American Philosophical Society Archives, Philadelphia, Pennsylvania.

[18]Quoted in Greville Bathe, *An Engineer's Miscellany* (Philadelphia, 1938), pages 122-123.

[19]*New-York Gazette,* 1 August 1774, in Rita Susswein Gottesman, editor, *The Arts and Crafts in New York, 1726-1776* . . . (New York, 1938), pages 319-320; Edward Wegmann, *The Water-Supply of the City of New York, 1658-1895* (New York, 1896), page 4.

[20](Purdie) *Virginia Gazette,* 3 March 1775.

[21][Bangs], "1776," *Historical Magazine,* volume 4 (1868), pages 305-306.

[22]J. Leander Bishop, *A History of American Manufactures from 1608 to 1860.* Volume 1, page 502; James B. Hedges, *The Browns of Providence Plantations: Colonial Years* (Cambridge, 1952), page 139.

[23]Hedges, *Browns,* pages 270, 313.

[24]William Parker Cutler and Julia Perkins Cutler, editors, *Life, Journals, and Correspondence of Rev. Manasseh Cutler, LL.D.* (Cincinnati, 1888), volume 1, pages 205-207.

[25]Thomas Green Fessenden, *The Register of Arts, or A Compendious View of Some of the Most Useful Modern Discoveries and Inventions* (Philadelphia, 1808), page 298.

Chapter 2

Early Inventors and Experiments

I know of nothing so perplexing
and vexatious to a man of feeling
as a turbulent Wife and
Steamboat building.

JOHN FITCH[1]

When the British colonists came to settle America in the early seventeenth century, they left a pastoral country which had not yet suffered the throes of the Industrial Revolution. By 1750 that economic upheaval had begun in England but had scarcely touched the American shores. As the colonists became estranged politically in the decades that preceded the Revolutionary War, they sought to bolster their growing nationalism through economic self-development. This growth took two forms: the fostering of manufactures and the improvement of transportation. Begun before 1776, both movements were disrupted by the war. After the cessation of hostilities and with the birth of the new Confederation, minds turned again to the creation of an American empire, based on western lands, mercantile legislation, and a republican form of government.

England was ever the prime example of the successful empire. During the latter half of the eighteenth century, British entrepreneurs were busy digging canals, mining coal, mechanizing the textile industry, erecting steam engines, and generally laying the groundwork for a century of industrial dominance. Similar activities were taking place in America, on a smaller scale. In 1790 Samuel Slater arrived with his knowledge of the new textile machines. Work was begun or renewed on a dozen canals along the Atlantic seaboard. Societies for the encouragement of domestic manufactures sprang up in the cities of the new nation. Mechanics and artisans found themselves in great demand; inventors dunned the various states and the federal government with petitions for patents.

The encouragement of manufactures and the improvement of transportation went hand in hand, and both were essential to the establishment of a balanced, integrated, and self-sufficient economy. In America, however, the problem of transportation seemed most urgent. Water-mill sites were plentiful, though not of equal worth, and small mills for producing such items as flour, lumber, paper, and fulled cloth had been common enough in colonial days. Such activities could be extended almost at will. Adequate transportation presented more difficult problems. The heavy and clumsy Newcomen engine, with its vast brickwork and ponderous beam, seemed to offer no solution to these problems. The Watt engine, developed shortly before the Revolution, was quite another story. The export of steam engines from England was banned in 1785, and even detailed information about them was not widely available. Their effect on English industry had been profound, but they had not yet been applied to transportation which, in England, was being adequately supplied at that time by newly built canals.

In 1764 James Watt, instrument maker at Edinburgh University, received a demonstration model of the Newcomen engine for repair. By the summer of 1765, he had improved the engine by adding a separate condenser. From 1765 to 1774, he was able to devote only about three years to the improvement of the steam engine. Coming then under the patronage of the successful manufacturer Matthew Boulton, he was able to concentrate his work and in March 1776 set up his first two commercially built engines, a 50-inch water-pumping engine for the Bloomfield Colliery and a 38-inch engine to provide the air blast for the furnace of the prominent iron maker, John Wilkinson.

For the five years between 1775 and 1780 Watt kept busy designing individual engines to customer specifications. Each engine represented an application of his established principles, but the various engine sizes and uses were worked out in each case. In 1781 Watt began again to make major improvements in the engine and produced, within a comparatively short time, devices for converting the engine's reciprocating motion into rotary, the double-acting engine, the method of using expansive steam an automatic centrifugal governor, and a compound (two-cylinder) engine.

Figure 2 represents a low-pressure Watt engine as it was built about 1800. On the downward stroke (diagram II, broken lines), steam from the boiler (B) entered the cylinder (C) above the piston (D), exerting a downward pressure. At the same time the steam below the piston was allowed to exhaust into the separate condenser (A), where a jet of cold water caused the steam to condense, creating a vacuum in both the condenser and the cylinder (below the piston), as well as in the pipe between. The pressure above the piston, combined with the vacuum below, caused the piston to descend. For the upward stroke (diagram II, solid line), valves reversed the flow of both the live and spent steam.

Diagram I represents the same engine, showing Watt's major contributions. His great improvements in the Newcomen engine included the application of a device for changing the reciprocating motion of the engine into a rotary one. Shown here is a simple crank (G) which was the easiest and best device for the purpose, although to avoid the possibility of infringing patents, Watt for some time used a

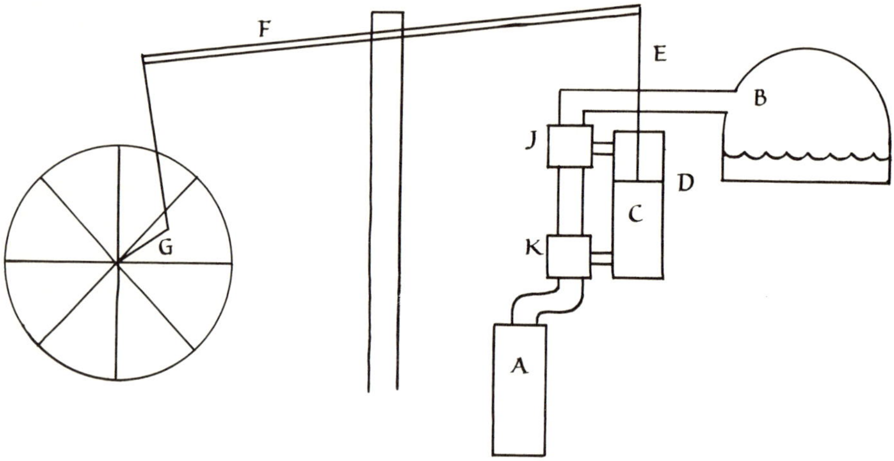

II

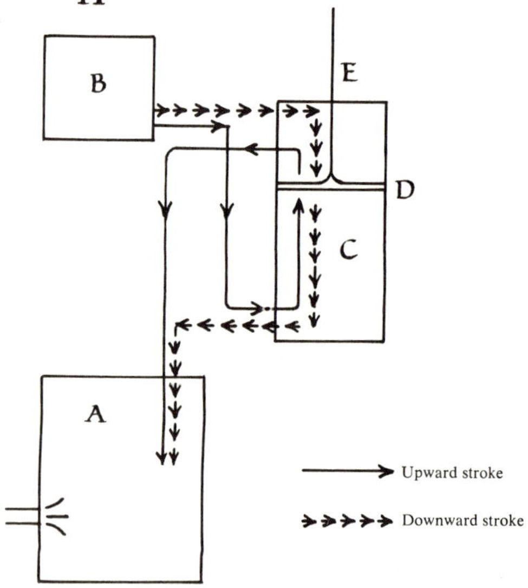

A Condenser
B Boiler
C Cylinder
D Piston
E Piston rod
F Beam
G Crank
J Valve
K Valve

Figure 2 Watt engine, about 1800

set of gear wheels, which he called the sun-and-planet, to accomplish this task. A second device was the separate condenser (A), by which he was able to maintain high cylinder temperatures and thus effect a saving of that fuel which had previously been used in reheating the cylinder after each condensation. Third, by enclosing the top of the cylinder and applying steam to each side of the piston alternately, he was able to increase the engine speed.

The application of these improvements in the steam engine–improvements which made it for the first time of potential interest to American entrepreneurs–took place between 1776, when Watt set up his first commercial engines, and 1786, when his double-acting, rotative engine made its public appearance at the Albion Mills in London. During this same period, the American colonies were transforming themselves into a new nation. This was a period of grand dreams but one during which high considerations of state and low problems of survival were dominant. Although such wartime activity as the letting of contracts for the casting of cannon no doubt added to the new nation's industrial experience and capability, the great industrial changes taking place in England were not quickly or thoroughly known to Americans.

Whatever knowledge of the Watt improvements had made its way across the Atlantic during the war years, its currency was not wide and many persons who might be expected to know about the improvements did not. Thomas Jefferson, for example, had long had a wide and deep interest in both technology and philosophy, but during the war he was much too busy to concern himself overly with either. When he replaced Franklin as American minister to France in 1785, he apparently had no knowledge of Watt's recent developments. In reply to a question from his old teacher, the Reverend James Madison, addressed to him in Paris, Jefferson replied:

> You . . . speak of a new method of raising water by steam which you suppose will come into general use. I know of no new method of that kind and suppose (as you say that the account you have received of it is very imperfect) that some person has represented to you as new a fire engine erected at Paris and which supplies the greater part of the town with water. But this is nothing more than the fire engine you have seen described in the books of Hydraulics . . . the idea of which was first taken from Papin's digester.[2]

Six months later, Jefferson traveled to England and was both surprised and excited by the industrial activity he encountered. "I could write you volumes on the improvements which I find made and making in the arts," he wrote enthusiastically. "One deserves particular notice," he added, "because it is simple, great, and likely to have extensive consequences. It is the application of steam for working grist mills. I have visited one lately made here."[3] This establishment was the Albion Mills, and in a letter later that year, Jefferson told Thomson that he had had a talk with "the famous Boulton" and was able to confirm that the engine at the mills was superior to the old type both in its rotary motion and in its greater economy.[4]

As the first city of the empire, London not only consumed a great quantity of flour annually but also exported a large amount. By the 1780s this need for flour was supplied by the products of some five hundred small gristmills, powered by wind or water and operating in the general area of the city. These mills were all small, representing an annual value of some £200 each and working only three sets of stones. In 1782 Samuel Wyatt began to plan a large gristmill, to be valued at £3000 per annum and to run a large number of stones. A site was selected at Blackfriar's Bridge on the Thames River in London, and Boulton and Watt were asked to provide an engine for the mill. Because a successful establishment would serve as a model and advertisement for the excellence of their newly developed double-acting, rotative engine, the partners joined the milling company and agreed to provide the necessary engines.

John Rennie, a young Scot engineer just then at the beginning of his brilliant career, was placed in charge of setting up the engine, a 50-horsepower model which was put in motion in 1786 and eventually operated eight pairs of stones. After some initial difficulties the mill performed well, and a second engine of the same size was erected in 1788. Poor management made the enterprise unprofitable, and when the mill burned in 1791 it was operating at a considerable loss. When Jefferson saw the mill in 1786, however, the technical success was assured and the economic failure not yet apparent. This first steam-driven mill, which Jefferson was able to inspect personally, was described for the American reading public in the pages of the *Columbian Magazine* in July 1788. It was a wonder of the industrial world, and its lesson was not lost upon would-be projectors—English and American.

As postwar contact with England increased, more and more Ameicans became familiar with the potential of steam power. John Noyes returned to Massachusetts with a letter from John Adams, who was then American minister to the Court of St. James. He petitioned his state legislature for a monopoly on the application of steam power to the manufacture of iron. He and his partner, Paul Revere, were granted the privilege.[5]

The seeking of government support for the introduction of new technologies was a natural prelude to enterprise in a century when patronage, in many fields, was so often the key to successful operation. In the spring of 1784 Silas Deane, the former Connecticut merchant and discredited American diplomat, wrote to his brother from England that he had just "made a Tour through all the manufacturing Towns, of this Kingdom, & . . . examined their late, & new inventions, & Machines of various kind." He had, he said, "seen Mills, on a New, and different Construction and every way preferable [to the old style], they are both Corn Mills, & Iron Works, rolling & Slitting Mills worked by a Steam Engine which carries them with an amazing degree of Force, and without any greater interruption or cessation than what You choose."[6]

Deane hoped to introduce the Watt engine into the United States but received divided counsel. A business associate advised him that "the Machine you mention must be very useful, but it will not be wanted for any thing of consequence, besides Pumps, for some time to come." Nevertheless, he concluded, "you would find no difficulty in obtaining a Patent here, could you find some new Invention that would be profitable & I have no doubt in your Tours, you must have discov'd many."[7] It was a plan which Deane decided to follow. In the spring of 1785 he wrote to his brother from Castle Head in Lancashire, where he was "with the Gentleman who makes the engines." With a boldness typical of the period, he revealed a plan to obtain patents of monopoly from Virginia, North Carolina, Maryland, Connecticut, New York, New Jersey, and Pennsylvannia for

> using the new invented Steam engine in Corn, Tobacco, & Saw Mills for a certain Term 14 or 16 Years, at least. My reasons for this, are that Our Funds being small, We must begin in a stile proportionate to Our Funds, and As soon as We have begun, larger capitalists will come in, & take the Business in effect from Us, or at least throw Us into the shade, whereas on the contrary Patents from all the principal States, will secure it to Us, & on the Strength of them I can take in Partners here [in England], by whose assistance a Capital To any Amount may be raised, & the Business carried on with vigor, & to a large extent.[8]

In October his brother Barnabas Deane wrote from Hartford, Connecticut, that "I have now Laid in a Petition for a Patent for Twenty One Years to Our Assembly for the Liberty of Fire Engines for Grinding Corn & Sawing Timber."[9] Speaking presumably for his brother, Barnabas told the legislature that "your petitioner, for valuable considerations given to certain ingenious persons in Great Britain, the first and original inventors of the new improved steam engine . . . your petitioner has acquired a perfect knowledge of the construction thereof, with new improvements, and the application thereof to a great variety of important purposes; such as the raising of water, the manufacture of iron, and to the working of corn mills, saw mills, and of mills of every kind." The petition was not granted, but Deane refused to give up the idea of transplanting recent English technology, and the steam engine in particular, to American soil.[10]

Early in 1789, not long before his death, Deane wrote again on the subject. His correspondent, however, answered that the times were not propitious in America. George Washington had just been inaugurated as the first President of the United States, and the first Congress had been sitting less than a month. The use of steam engines for operating mills "must turn out profitable in a Country that has a fixed & Settled Government," he told Deane, "but I am sorry to say this is not the case at present in any of the States of America, the people at large are dissatisfied, and a change from the present or a total revolution must take place before any plans of Business can be adopted with safety."[11] The success of the new Constitution did not come soon enough for Deane, who died in August.

A different case was that of William Tatham. Born in England in 1752, Tatham came to this country in 1769, then returned to England in 1796. The following year he made inquiries of Boulton and Watt about exporting engines to the United States. We have no record of what use he intended for these engines. Since he was a civil engineer, however, it is possible that he wanted either to pump water for canals or perhaps to apply engines to boats. Like many such plans, his came to nothing. He wrote tracts on canal transportation and in 1801 was made superintendent of the London docks. He returned to the United States in 1805 and died in Virginia in 1819, but there is no evidence that he interested himself again in steam engineering.[12]

While Jefferson was in Europe catching up with the latest advances in Old World technology, his American correspondents kept him abreast of New World developments. He wrote to Charles Thomson in 1786, "I heard you are applying . . . [steam] in America to navigate boats, and I have little doubt but that it will be applied generally to machines, so as to supersede the use of water ponds, and of course to lay open all the streams for navigation. We know that steam is one of the most powerful engines we can employ; and in America fuel is abundant."[13] The following year Francis Hopkinson wrote Jefferson that "as to Philosophy–We abound in Schemers and Projectors. There is one *Fitch* who has been the Twelve month endeavouring to make a Boat go forward with Oars, worked by a Steam Engine." "I have no Doubt," he added, "but that a boat may be urged forward by such Means, but the enormous Expense and Complexity of the Machine must prevent its coming into common use."[14]

Jefferson was closer to being correct than was Hopkinson. While the old-style Newcomen steam engine had little relevance to the economic needs of the new American nation, the Watt engine–more powerful, more efficient, and equipped with rotary motion–was immediately seen by many as the solution to what was probably the nation's major economic problem, adequate transportation. During the next decade and a half before the turn of the century, inventors from Massachusetts to Georgia and from New Jersey to Kentucky were attempting to build steamboats. From encyclopedias and talks with recent emigrants from England, from experience with the Hornblower, Colles, and Brown engines, and from news of the efforts of others, these mechanics built steamboats that actually worked.

John Fitch was born in Connecticut in 1743, and during the next forty years suffered through an unhappy marriage, an Indian captivity, unsuccessful land speculations, and undistinguished careers as silversmith, clockmaker, and surveyor. Then in April 1785, while walking along a road in Bucks County, Pennsylvania, he was "so unfortunate . . . as to have an idea that a carriage might be carried by the force of steam along roads. I pursued that idea about one week," he later wrote, "and gave it over as impracticable, or, in other words, turned by thoughts to vessels."[15] The story has the ring of myth, and it is probably safe to assume that Fitch had previously seen the force of steam demonstrated in some form. He is quite explicit in stating that he had at that time never seen a steam engine, nor did he know that such a machine had been invented. In this he was quickly set straight for when he took preliminary drawings to show his friend the Reverend Mr. Irwin, of Neshamony, the latter showed him a description of the Newcomen engine in a book from his library.

Fitch no doubt revised his plans in the light of this information and showed his new drawings to various men of scientific reputation in Pennsylvania for their approval. He also began immediately to seek the patronage of various governments, among others that of the Spanish king. Clearly he was already considering the vast potentialities of steamboats on the western (and still Spanish) waters. In October 1785 Fitch made a trip to Kentucky in connection with his land speculations and on his way stopped at Lancaster, Pennsylvania, to speak with William Henry.

Henry had been trained as a gunsmith and had become one of the most prominent citizens of the inland manufacturing city. When the Continental Congress was forced to flee to Lancaster in 1775, Henry entertained its members and talked about steam with such leading figures as Thomas Paine, David Rittenhouse, and Andrew Ellicott. When Fitch visited Henry in 1785, it seems likely that he met with more enthusiasm than practical aid, although Henry may have been the one who encouraged Fitch to submit his proposal to the American Philosophical Society in December 1785.[16]

Early in his work Fitch planned to order an engine from England—as did Robert Fulton two decades later—and his fate might have been different if he had been able to do so. In a petition to the Maryland State Assembly in January 1786, Fitch announced that he had plans for "a Machines for promoting Navigation" and "proposes to send to Europe for a Steam Engine to Effect it. He prays this Honourable Legislature to lend a sum sufficient for that purpose and impower Andrew Ellicotte Esqr, under whose Patronage he wishes to erect it to draw out of the Treasury a sufficient sum to procure Sd. Engine and the subscriber will refund the same" from expected profits on a map of the western country which he had drawn up.[17] The Maryland Assembly denied his petition and Fitch was forced to change his plans. He apparently abandoned hope of securing an English engine at this time and turned to the idea of building one himself.

Armed with grants of monopoly from several states, Fitch formed a company of some forty subscribers in April 1786 to build and operate his steamboat. Among the company was William Thornton, recently removed from the West Indies and holder of a medical degree from Edinburgh. The presence of Thornton in the group was fortunate, for he told Fitch about James Watt's recent improvements and suggested that the proposed steam engine be built to incorporate them. In later years Thornton wrote that "Watt & Boulton's Steam Engine was recommended by me to John Fitch's Steam Boat Company, when the first Boat was built . . . & we made one on the principle of Watt & Bolton's Engine." Furthermore, he asserted, "I believe I was the only person in the Company who had ever seen a Steam Engine, but those I saw when very young & they only worked one way. We tried various modes and made an Engine work both ways."[18] If Thornton's foresight was as shrewd as his hindsight, his knowledge was one of the most valuable assets of the ill-fated company.

Thus strengthened with the backing of a respectable if not vigorous company and with some idea of what he sought, Fitch asked John Nancarrow to draw up plans for an engine. Nancarrow operated a steel furnace in Philadelphia, and was described by his partner as "an Engineer by profession and much accustomed to the erecting of steam engines in England."[19] Fitch had by this time taken Henry Voigt, an ingenious clockmaker, as his chief mechanic; the latter showed Nancarrow and his partner Timothy Matlack his newly invented spiral-tube boiler, intended to be used on Fitch's boat. Both consultants advised ' Mr. Voight to exhibit the first experiment on the steam boat, by means of the common great [grate] boiler, rather than involve himself in the difficulties and delays which would probably arise in the execution of his new mode of generating steam."[20]

Nancarrow's advice to Fitch on his proposed steam engine was similarly conservative. In May 1786 Fitch wrote that he "had the Promise of Mr. Nancarrow to give me all the assistance that he could, and by relying on him have begun my Works."[21] As Nancarrow's work progressed, however, it became clear that he intended to build the Newcomen type engine with which he had been familiar in England. Fitch complained that "it was to work upon the old-fashioned plan of engine, and was to have a weight to raise the piston."[22] This disagreement was instrumental in halting Nancarrow's participation in the plans.

Fitch turned next to the best known steam-engine builders in America: Josiah Hornblower and Christopher Colles. He explained to them that he was "indeavouring to impel a Boat forward by the force of Steam." He confessed that although Nancarrow had agreed to help him with the works, "Now I find I am under the necessity of Compleating them my self, and being an intire stranger to Steam Engines I humbly request your opinions on the following Plan." The description which followed revealed his plan to build a Boulton and Watt engine, differing mainly in having its cylinder placed horizontally. The bulk of his description concerned the justification for and construction of a separate condenser. He requested Colles and Hornblower to "investigate this draft and point out my errors or inform me where I am likely to meet with the most difficulty or whether or not the present plan may succeed."[23]

Hornblower, then serving in Congress in New York, answered Fitch that "I have examined your plan for propelling a Boat forward by the force of a Steam Engine, and do apprehend it to be rather too complicated to succeed well in practice." He registered strong objections to the horizontal position of the cylinder ("it will be apt to make a leaky Piston") and the use of the condenser and air pump for evacuating it, but ended with wishes for success.[24] As Fitch discovered, the experts advised the path of tradition rather than innovation–a tradition more irrelevant than hostile.

Thrown largely on his own resources, Fitch found that his self-confessed ignorance was his greatest handicap in building a successful steamboat. The harshest critic of Fitch's mechanical abilities was his own partner, Henry Voigt. Writing many years later to Thornton, Voight asserted: "that the Steam boat company miscarried . . . was entirely owing to [Fitch's] . . . want of Mechanism; he was the most deficient in that respect of any man I ever know that professed it. Every thing that was done in that line," he continued, "was with him either too good, or good enough; because he was no more a judge of those matter than the person who executed it; nor could he make any mechanic understand what he wanted; and I was sure he would fail."[25] Indeed his complete lack of mechanical knowledge or aptitude cast an aura of confusion and frustration over his entire career and colored both his few successes and his many failures.

Fitch was aware of his ignorance, but was confident of his ability to learn. He told the Pennsylvania legislature, "Mr. Voight and myself are sure that we can build an Engine; nay, we are vain enough to believe that we can make one as good as they can in Europe."[26] But after five years of constant struggle, he admitted that "I had no conception . . . that it would have employed so great a length of time merely to

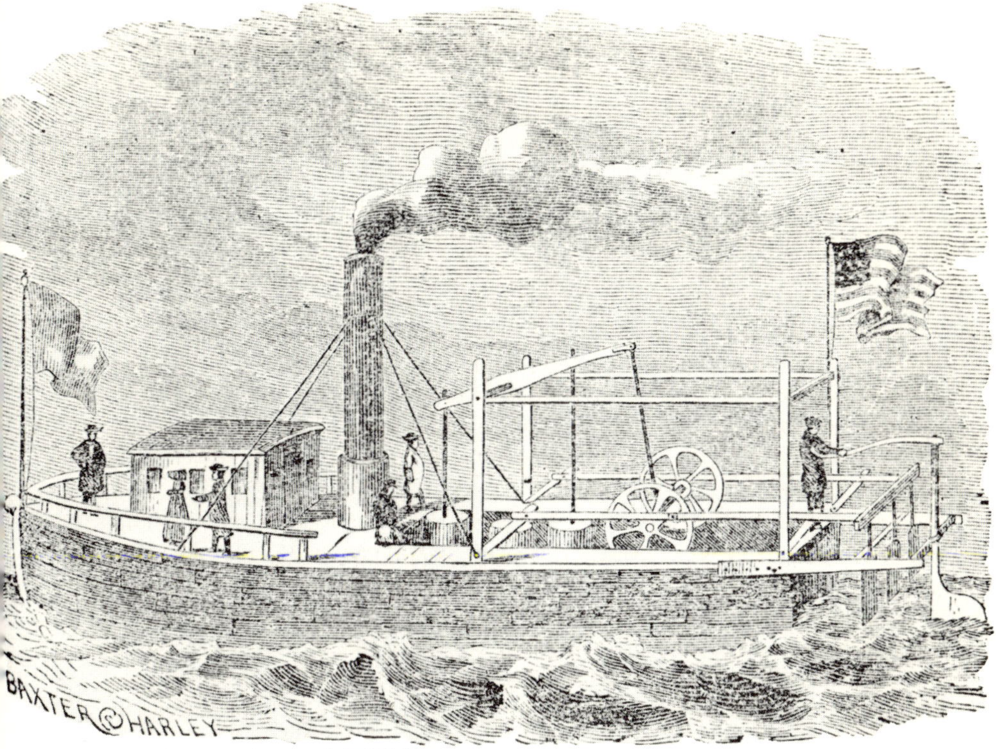

John Fitch's steamboat of 1788-90.
From Thompson Westcott, *The Life of John Fitch, the Inventor of the Steamboat* (Philadclphia, 1857), page 284.

acquire the art of making a steam engine—an art familiar in England, but of which all the artizans I could ever converse with in America, [were] entirely ignorant. Thus we had to explore an unbeaten path, and did not ascertain the true course until we had wandered into a thousand wrong Rodes."[27]

His workmen's ignorance of steam engines was another of Fitch's major problems. These early American experimenters were constantly frustrated by the necessity of actually subsidizing the education of their workmen. Thornton claimed that "we made the whole Engine, & even formed the Engineers out of common Black Smiths not one of whom had ever seen an Engine."[28] Eventually these men did learn something of their new trade, and Fitch reminded Congress that "if we never carry it to any greater degree of perfection, we have merited a generous reward for laying the foundations for future improvements. We have by this means introduced Engineers into our Country, and Consequently, one of the first powers of Nature into our Empire, which may be useful in most great works."[29]

What Fitch apparently never realized was that the problem of building steam engines went much deeper than mere lack of knowledge or experience. He faced a fundamental lack of industrial capacity as well. A first-rate mechanic with first-rate tools and facilities could have done the job, but such a person never came within Fitch's operations. The simple but critical demands of proper metal casting, machining within acceptable tolerances (and according to Voigt, Fitch had a great deal of tolerance indeed), and fabricating with materials of sufficient strength proved beyond the capacity of his associates. The fact that Fitch never really appreciated the importance of these factors only made the frustration more certain and bitter.

Fitch claimed only the distinction of being the first to apply steam to moving boats and denied his priority to either the engine itself or its application to mills or mines.[30] He did, however, foresee the future usefulness of the engines to American manufactures and pointed out that through his efforts to develop the steam engine, he had paved the way for the introduction of industry into those parts of the country unsuited to water mills. Such enterprise as his, he noted, had been patronized by both Louis XVI and "the Potent Empress of Russia." Surely then, he argued, should not the new American republic show at least an equal amount of enlightenment?[31]

The rest of Fitch's story has earned him the well-deserved appellation "Poor Johnny" Fitch. After successfully operating a boat between Philadelphia and Burlington, New Jersey, during the summer of 1790, he sought financial support in France, England, and again in the Court of Spain. Finally despairing of patronage, he settled in Kentucky, dying there by his own hand in 1798.

The ultimate failure of John Fitch is instructive not only because of the advantages he won but also for the handicaps under which he suffered. His real assets were not negligible. He conducted his experiments in Philadelphia, which was certainly the seat of the mechanic arts in this country and the center of finance as well. He was successful in forming a respectable company that was composed of men possessing reputation and substance. He enjoyed the counsel and advice of such men as Oliver Evans and John Nancarrow, and the endorsement and encouragement of such scientific leaders as David Rittenhouse and Andrew Ellicott. Finally, he began his work at a time when America was seeking her economic independence with a patriotic fervor. This was a time of ferment, technological as well as political.

Great as were Fitch's advantages, his handicaps were even greater. Overriding among these was his total ignorance of steam technology. He had never seen an engine of any kind–nor had most of his men. Those who had seen an engine were ignorant of the details of the Watt improvements which were indispensable for his purposes. Conservative advice to hold to the tried and true Newcomen engine was worse than useless to a man who wanted, as Fitch said, to "produce a steam engine floating on the water."[32] Had he been able, like Robert Fulton, to buy a Watt engine in England and merely attach it to a boat, he might have been more successful. As it turned out, he wasted his substance and that of the company on fruitless attempts to build what might better have been purchased abroad. For example, at least seven different condensers were tried in his one really successful boat. Perhaps almost as great a handicap as his own ignorance was his need to let out various components of his machinery to various makers whom he could not easily supervise and the quality of whose work he could not control. When success came to other men in later years, it was generally to those who were able to operate from an ironworks of their own, where instructions were followed and standards were set and maintained.

During the two decades following Fitch's attempt to perfect the steamboat, at least a dozen other American experimenters tried their hand at the same problem. Most of them had little success and left no record of their mechanisms. Such men as Jehosaphat Starr and Apollos Kinsley labored with little publicity and less success. Others received more contemporary notoriety or scientific comment. For example, in 1788 it was announced that William Longstreet and his partner Isaac Briggs had received a monopoly in Georgia for the use of their improved steam engine.[33]

This double-acting engine used a horizontal piston. The key improvement was believed to be the method of condensing steam to create a vacuum. The exact method was not clearly described but was said to substitute a metallic pipe filled with cold water at each end of the cylinder for the jet of water used in the Newcomen engine. This may have been a garbled attempt to reproduce Watt's improvements, or perhaps it was an independent effort to avoid the wasteful cooling of the entire cylinder and may have been an early surface condenser.[34] The inventors apparently did not build a boat in which to use their engine at this time.

Isaac Briggs dropped out of the enterprise, and in 1790 Longstreet tried independently to raise money to build a steamboat. Writing from Augusta, Georgia, in that year he confessed that "until very lately I did not think that either artists or material could be had in this place sufficient. However, necessity—that grand source of invention— has furnished me with an idea of perfecting my plan, almost entirely with wooden materials, and by such workmen as may be gotten here. . . ." This reliance on wood and local mechanical skill and facilities boded ill for the project, and the plan seems to have been dropped once again. It was not until 1807 that he returned to the problem of steamboats and finally produced a small one, propelled by long poles pushing against the river bottom. The engine cylinder was only four inches in diameter, but this experimental boat is reported to have run up the Savannah River at five miles an hour.[35]

Longstreet was not idle during those intervening years and began work on a cotton gin before 1790. As was probably inevitable, there were rumors that Eli Whitney had benefited from his early experiments. In 1792 Longstreet announced a new steam engine designed to turn sawmills and gristmills. Again the description was vague, but clearly indicated that this time the inventor was attempting to introduce a rotary engine, "without the complication of wheels, cogs, or cranks; and all its operations are managed by a single cock." With this or a more conventional engine, Longstreet established two cotton gins run by steam. Unfortunately, both were destroyed by fire. He also set up a steam gristmill near Saint Marys, in Georgia, but this was burned by British soldiers during the War of 1812. These setbacks exhausted Longstreet in spirit as well as fortune, and he died in 1814.[36]

Isaac Briggs is not mentioned after 1788 in connection with steam experiments, but his brother Samuel Briggs continued active in his own right. The latter turned some casting patterns for John Fitch in 1786 and probably learned something about steam engines from this experience.[37] With knowledge of Longstreet's efforts, perhaps from his brother Isaac, Samuel Briggs in 1802 patented a steam engine of the rotary type.[38] William Thornton was greatly impressed with this rotary engine and recommended it to a prospective steamboat builder.[39] Samuel Briggs also took out patents for steam engines in 1809 and 1812.[40]

Elijah Ormsbee, another early steamboat projector, worked as a carpenter at an iron mine at Cranston, Rhode Island. While working on the repair of the large Newcomen engine erected by Joseph Brown, Ormsbee met David Wilkinson, a young machinist and mechanic from Pawtucket. Ormsbee told Wilkinson that he had just been reading of a steamboat operating at Philadelphia (no doubt that of Fitch) and suggested that they build one of their own. Wilkinson returned home, made the patterns, cast the parts, bored the cylinder, and together the two men assembled a working steamboat.[41] Efforts on this boat were soon discontinued for lack of funds, but Ormsbee went on to erect an engine in Providence to pump water for a distillery. This engine did not work well and was abandoned within a year.[42]

An important result of the Ormsbee experiment was that it introduced Daniel French to steam engineering. While Wilkinson was busy on the Ormsbee boat, he was visited by young Daniel French from Connecticut who expressed an interest in learning how the vessel was to operate. French later had an important career in steam work.

This incident points up the vital importance for steam engineering in early America of the cross-fertilization of efforts through frequent interchange of men, inspiration, and ideas. Hornblower had brought the first steam engine to America. He advised John Fitch, whose efforts were publicized in the newspapers and read by Ormsbee, who was repairing the Brown engine in Rhode Island. There Daniel French saw the work in progress and later, about 1809, moved to New York where he became acquainted with Robert Fulton.[43] With his helper, Hezekiah Bliss, French built steam engines in New York and Philadelphia and built at Pittsburgh one of the first steamboats to ply western waters.[44] In his turn, Bliss moved forward to a colorful career as an Ohio riverboat captain and then, in partnership with Dr. Eliphalet Nott, established the Novelty Iron Works in 1827. By the middle of the nineteenth century, it was the largest manufacturer of steam engines in the city of New York and one of the largest in the entire country.

In spite of the many steamboat experiments, neither permanent establishment of such a public conveyance nor public acceptance of this mode of travel was accomplished until the success of Robert Fulton in 1807. But to conclude that the experiments led to no good whatever would be a great error. Many of the men who tried and failed before Fulton, became active again after the *Clermont* firmly established the steamboat in the public favor. Nicholas I. Roosevelt, Samuel Morey, Nathan Read, John Stevens, and others saw their early efforts vindicated and again turned to steam. Furthermore, some of the workmen who had labored on the earlier attempts were directly instrumental in assisting Fulton. As a group, these men were the first to struggle with the steam engine as it had come from the hands of James Watt.

The uniform failure of early steamboat projectors can be attributed to several circumstances. For one, they often lacked the financial backing and commercial sense which are so essential during long periods of development to insure economic success. On a technical level, the proper mode of propulsion—poles, paddle wheels, jets of water, screw propellers—was not obvious and its linkage to the engine was not simple. Finally—through ignorance, necessity, or choice—each of these men decided to build his own steam engine and some even went so far as to almost reinvent it. At the very least this necessitated the acquisition of special tools and skills, and beyond this it meant that the mechanical excellence of the engines could not rise above the technical resources of the immediate area. The problems of casting, boring, and machining metals were difficult enough to solve in the cities of New York and Philadelphia—in Cranston or Savannah they were virtually insurmountable. Between the ingenious conception and the economic need lay the critical factor of mechanical capability.

Notes

[1] Quoted in Westcott, *Fitch,* page 121.

[2] Thomas Jefferson to Rev. James Madison, 2 October 1785, *The Papers of Thomas Jefferson,* edited by Julian Boyd (Princeton, 1950-), volume 8, page 574.

[3] Thomas Jefferson to Charles Thomson, 22 April 1786, ibid., volume 9, page 400.

[4] Thomas Jefferson to Charles Thomson, 17 December 1786, *Writings of Thomas Jefferson,* edited by P. L. Ford (New York, 1892-1899), volume 4, pages 337, 340.

[5] Bishop, *American Manufactures,* volume 1, pages 498-499; Oscar and M. F. Handlin, *Commonwealth: Massachusetts, 1774-1861* (Cambridge, 1947), page 81.

[6] Silas Deane to Simeon Deane. London, 3 April 1784, in *Correspondence between Silas Deane and His Brothers and Their Business and Political Associates, 1771-1795,* Collections of the Connecticut Historical Society, volume 23 (Hartford, 1930), page 197.

[7] Jacob Sebor to Silas Deane, New London, 10 November 1784, ibid., pages 204-205.

[8] Silas Deane to Simeon Deane, Castle Head in Lancashire, endorsed 20 May 1785, ibid., page 211.

[9] Barnabas Deane to Silas Deane, Hartford, 15 October 1785, ibid., page 216.

[10] "Connecticut. Extracts from Colonial and State Papers, in the Office of the Secretary of State, Hartford," *Report of the Commissioner of Patents, for the Year 1850,* House Executive Document 32, 31st Congress, 2nd session, page 444; Bishop, *American Manufactures,* volume 1, page 521.

[11] Frederick William Geyer to Silas Deane, Boston, 1 May 1787, in Deane Papers, *Collections of the Connecticut Historical Society,* volume 23, page 244.

[12] Roll, *Early Experiment,* page 240; "William Tatham," *Appletons' Cyclopaedia of American Biography,* volume 6, page 37.

[13]Thomas Jefferson to Charles Thomson, 22 April 1786, *Papers of Thomas Jefferson* (Boyd), volume 9, page 401.

[14]Francis Hopkinson to Thomas Jefferson, 8 July 1787, ibid., volume 11, page 561.

[15]Quoted in Westcott, *Fitch,* pages 119-120.

[16]Francis Jordon, *The Life of William Henry of Lancaster, Pennsylvania, 1729-1786; Patriot, Military Officer, Inventor of the Steamboat,* pages 51-52. Thurston thinks that Fitch may have gotten his original idea from Henry. Robert H. Thurston, *A History of the Growth of the Steam-Engine,* page 231. It is also significant that Robert Fulton was a frequent guest at Henry's house. Ibid., page 231.

[17]Petition dated 5 January 1786, Fitch papers, volume 1, LC.

[18]"Answers by William Thornton . . . to numerous Interrogations," no date, William Thornton papers, volume 16, LC; William Thornton to Robert Fulton (?), 16 December 1807, Thornton papers, volume 3, LC.

[19]Testimony of Timothy Matlack, 14 July 1788, Fitch papers, volume 2, LC.

[20]Ibid.

[21]John Fitch to Josiah Hornblower and Christopher Colles, 17 May 1786, Fitch papers, volume 1, LC.

[22]Quoted in Westcott, *Fitch,* page 152.

[23]John Fitch to Josiah Hornblower and Christopher Colles, 17 May 1786, Fitch papers, volume 1, LC.

[24]Josiah Hornblower to John Fitch, 24 May 1786, Fitch papers, volume 1, LC.

[25]Henry Voigt to William Thornton, 29 June 1809, Thornton papers, volume 3, LC.

[26]Quoted in Westcott, *Fitch,* page 166.

[27]Quoted in ibid., page 306.

[28]William Thornton to John Stevens, 8 July 1809, Thornton papers, volume 3, LC.

[29]Quoted in Westcott, *Fitch,* page 201.

[30]Ibid., page 172.

[31]Ibid., pages 189, 187.

[32]Quoted in ibid., page 172.

[33]Victor S. Clark, *History of Manufactures in the United States* (New York, 1929), volume 1, page 49. Clark also says that they received a patent from South Carolina in 1786. For Longstreet, see sketches in *Dictionary of American Biography* (New York, 1928-1944), volume 11, page 393 and *Appletons' Cyclopaedia of American Biography,* volume 4, page 16. William Longstreet was the father of Augustus Baldwin Longstreet, noted humorist and author of *Georgia Scenes.*

[34]*Columbian Magazine,* volume 2 (March 1788), page 169; *Gazette of the State of Georgia* (Savannah), 13 March 1788.

[35]John Donald Wade, *Augustus Baldwin Longstreet: A Study of the Development of Culture in the South* (New York, 1924), pages 9-10.

[36]Ibid., pages 10-11; *Universal Asylum, and Columbian Magazine,* volume 9 (October, 1792), page 286.

[37]John Fitch, *The Original Steam-Boat Supported; or, A Reply to Mr. James Rumsey's Pamphlet* (Philadelphia, 1788), pages 17-18.

[38]Memo of William Thornton, 12 October 1802, Thornton papers, volume 3, LC.; *Subject-Matter Index of Patents for Inventions Issued by the United States Patent Office from 1790 to 1873, Inclusive* (Washington, D.C., 1874), volume 3, page 1427.

[39] William Thornton to Isaac Briggs, 26 October 1802, Thornton papers, volume 5, LC.

[40] Westcott, *Fitch,* page 406fn; *Index of Patents,* volume 3, page 1427. For specifications of Samuel Briggs' various engines, see Isaac Briggs papers, LC.

[41] "David Wilkinson's Reminiscences," *Transactions of the Rhode Island Society for the Encouragement of Domestic Industry. In the Year 1861* (Providence, 1862), page 104. For Wilkinson's career as a machinist, see Jonathan Thayer Lincoln, "The Beginnings of the Machine Age in New England: David Wilkinson of Pawtucket," *New England Quarterly,* volume 6 (December 1933), pages 716-732.

[42] "Statement of Capt. John H. Ormsbee," *Transactions of the Rhode Island Society for the Encouragement of Domestic Industry . . . 1861,* page 124.

[43] About 1840, a chance acquaintance told Wilkinson that Fulton had kept French "locked up for six months, making drafts for him." "Wilkinson's Reminiscences," *Transactions of the Rhode Island Society for the Encouragement of Domestic Industry . . . 1861,* page 105.

[44] Daniel French to William Thornton, 12 February 1814, Thornton papers, volume 4, LC. Cf. Louis C. Hunter, *Steamboats on the Western Rivers: An Economic and Technological History* (Cambridge, 1949), pages 136-137.

Chapter 3

Emigrant Engineers

To procure all such machines as
are known in any part of Europe,
can only require a proper provision
and due pains.

ALEXANDER HAMILTON[1]

Most of the earliest American engine builders and steamboat experimenters either were themselves emigrant English engineers with previous experience in steam technology, or their efforts were aided by such men. Whether they went on eventually to set up their own shops, dropped out of business completely, or continued in a supervisory capacity, these men were a major factor in the introduction of steam engines into the United States. As Alexander Hamilton noted in 1791, "ingenious and valuable workmen . . . by expatriating from Europe, have improved their own conditions, and added to the industry and wealth of the United States."[2]

Josiah Hornblower continued to be involved in early steam engineering in America. The large Newcomen engine which he erected at the Schuyler mine had burned down before the Revolution, but its influence continued to be felt. Nicholas I. Roosevelt and his associates leased the idle copper mine in February 1793. As Roosevelt went about restoring the mine and machinery, he came across some parts of an old "Atmospheric Steam Engine." He proceeded to rebuild the enginehouse and hired Hornblower to repair the engine.[3]

While Hornblower was thus engaged, Roosevelt hired workmen, principally miners, from England and Germany to operate the works.[4] He bought an additional tract of ground from Hornblower and constructed a foundry, machine shop, and stamping mill. The whole establishment was named Soho, after the Boulton and Watt Works in England. Hornblower soon had the engine working again, and it was able to drain the mines within four days. Roosevelt, however, was not satisfied with the engine, probably because it was still operating on the Newcomen principles.[5] At this time he collected about him a remarkable group of workmen, several of whom played major roles in early American steam engineering.

One of this group was John Hewitt, whose son Abram later became famous as an ironmaker and mayor of New York City. Hewitt, born in England in 1777, was trained in the engine works of Boulton and Watt, the former being a distant relative. He arrived in this country in January 1796 and began work at Roosevelt's Soho Works as a patternmaker.[6] He continued in Roosevelt's employ until 1802, when he turned to the trade of cabinetmaking.

Another of the group was Frederick Rhode, probably from Germany, who was listed in 1797 as assayer at the Schuyler mine.[7] Roosevelt later recalled that he placed George and Lewis Rhode, who were then boys, under one of his engineers, and that in the course of a few years they became not only acquainted with the principles but could also make an engine. Lewis Rhode became a trusted representative of Robert Fulton, and provided him with technical advice in his many law suits. He was killed in the machinery of a Brooklyn ferry in 1814, while on duty for the North River Steamboat Company. One of the Rhodes, possibly Frederick, set up a foundry in New York City and supervised the apprenticeship of James P. Allaire.

The Soho engineer who taught the young Rhode brothers was Charles Stoudinger, who came to New Jersey directly from college in Germany. He learned steam engineering at the Soho Works, and Roosevelt commented that "in a few years of theory and practice [he] became one of the first engineers in America."[8] The career of Stoudinger was stormy—mostly in the employ of Robert Fulton. He acted as his trusted agent and engineer and in 1810 Stoudinger represented Fulton in Pittsburgh when the latter's company built the first steamboat on the western waters. So close was his association with Fulton that Benjamin Henry Latrobe, after roundly condemning his "obstinancy, his ignorance, his eternal slowness, his hereditary & unimprovable stupidity," went on to say that "Fulton thinks him useful. I think him Fulton's spy."[9]

His later employment by Fulton also rankled Roosevelt, who wrote:

> Mr. Stoundinger entered my employ on his first arrival in America, when quite young; I first had him instructed by Mr. Smallman in the art of building steam engines and then, that he might have practical information, I employed him in various undertakings in that line, in which he has cost me many thousand dollars, as all his experience before he became perfect in his profession, was gained at my expense. When he entered the service of Mr. Fulton he was a first rate engineer, and he has no doubt contributed greatly to the success of Mr. Fulton's Steam boats . . . .[10]

Stoudinger was later honored by having at least two important steamboats named after him.

As Roosevelt indicated, Stoudinger's tutor at the Soho Works was James Smallman, chief among the former's workmen and later an engine builder in his own shops. It was to Smallman that Roosevelt turned when he became dissatisfied with the Newcomen engine which Hornblower had rebuilt. He had arrived late in America, after having filled responsible positions with the firm of Boulton and Watt. In Holland, he had been their agent in erecting an engine to drain a lake, and he had helped install the giant Watt engine in John Wilkinson's rolling mill at Bradley.[11] Such a well qualified engineer was not met with everyday, and after the Hornblower engine had been "improved," Roosevelt continued to employ him for a number of years "at a goodly salary."[12] It was with this group of workmen—Smallman, Stoudinger, Rhode, and Hewitt—that Roosevelt conducted his unsuccessful steamboat experiments in 1798.

Roosevelt had been driven from New York by the invading British Army during the Revolution. While rusticating on the Hudson, he had experimented with a small boat that was driven by paddle wheels powered by a hickory and whalebone spring. Upon returning to New York City, he conceived the idea of steam power for his paddle wheels. Sidetracked temporarily by his copper-mining venture, he returned to the steamboat idea late in the 1790s. He was then joined by John Stevens of Hoboken. They announced their intention to navigate by steam, and Robert R. Livingston was invited to join the partnership.[13]

Livingston, soon to be named American minister to France, had a fondness for dabbling in novelties. Not entirely unsuited to the role of gentlemanly amateur, nevertheless he seemed too much a promoter to suit his position as Hudson River patroon. Fortunately for his later patronage of Robert Fulton, his early partnership with Nicholas Roosevelt served more to educate than to discourage.

Livingston, too, had been thinking of building a steamboat. In December 1797 he disclosed his intention to Roosevelt, adding that "I was about writing to England for a steam machine but hearing of your wish I am willing to treat with you."[14] An agreement was reached and the works were begun. Smallman and Stoudinger contributed most of the skill; Livingston and Roosevelt provided the funds, facilities, and plans. A quarrel developed almost immediately, Livingston arguing that the power should be applied through a jet of water forced out the stern, driving the boat forward by reaction. Stevens was inclined toward the use of a screw propeller, and Roosevelt held out for paddle wheels hung over the sides. Livingston, "being anxious to be considered the patron of the arts and an universal genius," as Roosevelt commented sarcastically, won the battle but lost the war. The boat ran, but poorly.[15]

Smallman and the other Soho mechanics were convinced that the failure resulted from Livingston's method of propulsion.[16] Livingston thought the trouble lay with the engine. Roosevelt described the engine as being "upon Watt & Boltons plan, with this difference that I used a double air pump and wooden boiler."[17] Livingston did not look with favor upon the "improvements" made by Roosevelt. "I am more mortified than surprised at the failure of the Engine," he scolded, "I have long suspected that the pretended improvement on Doc. Watt would turn out as it has."[18] His dependence upon Roosevelt having led him to failure, he turned again to the idea of procuring an engine from England. He wrote James Watt asking for terms and confessed that "tho' steam engines are perfectly made here, yet the small number of workmen that understand the business renders it a slow and expensive operation."[19]

There can be no doubt that both the steam engine and the means of propulsion contributed to the failure of the Livingston-Roosevelt steamboat. The main result was that Roosevelt had once again helped finance the success of both Fulton and Stevens. When Fulton suggested buying the *Clermont* engine in England–rather than trying to have it made in the United States–Livingston was already convinced.

The next major undertaking of the Soho group was the construction of two large, Watt-type engines for use in the Philadelphia Waterworks. Contracted for in 1799, they represented the most ambitious use of steam yet attempted in this country. These engines were part of a plan developed by the young architect-engineer Benjamin Henry Latrobe to furnish water for the city. Born in England in 1764 and there trained as an architect, he came to Norfolk, Virginia, in 1796. Two years later he moved to Philadelphia, where he began a lifelong interest in the development of steam power.

Repeated epidemics of yellow fever had plagued Philadelphia and it was believed that noxious drinking water brought up from shallow wells which dotted the city, contributed to the general affliction. Latrobe suggested that two large steam engines be used to raise water from the Schuylkill River to a reservoir whence it could be distributed to the citizens through a system of pipes.

This well-publicized plan was ill received. Some opposition came from the supporters of the Schuylkill and Delaware Canal Company which hoped to supply water to the city as a part of its integrated plan of river development. More specific criticism came from those who were opposed to the use of steam in any way. Summarizing their arguments in 1811, James Mease claimed that they arose "from the novelty of the undertaking, the general failure of all previous attempts at the application of steam to mechanical purposes, in the United States, especially to boats: and our inexperience in the conducting of the operations of a steam engine on land, also contributed, in the minds of many of the citizens, to raise doubts of the successful issue of the proposed scheme."[20] One disgruntled citizen complained that "a fire engine to raise water sufficient for the city will consume an immense quantity of wood or coals," and added that "fuel seems already sufficiently high, without increasing it by such a scheme as this."[21]

Latrobe's defense of the scheme was largely to deny its weaknesses and to emphasize the great power and wonder of steam. The waterworks was to be his greatest enterprise and certainly one of the major engineering feats of the country. He was thinking in grand terms, and for all its practical justifications, the use of steam somehow fit the heroic mold of the whole project. "In every species of machinery," he argued, "in which mechanical powers alone operate, the bulk, the friction, and the unwieldiness of the works, encrease nearly in proportion to the effect required;–in the chemical operation of a Steam-engine, power is encreased in a ratio far outstripping the bulk and price of the engine." What he had to battle was the simple fact that while quantitative changes are easy enough to grasp, qualitative changes are sometimes singularly obscure. The matter of alleged inexperience was disposed of more quickly by affirming his belief that "this city can produce Smiths capable of constructing very efficient Engines, under proper direction."[22]

Latrobe won and in March 1799 the city of Philadelphia contracted with Nicholas Roosevelt for two steam engines, each capable of raising, per day, three million gallons of water fifty feet. Latrobe was retained as chief engineer for the works. Construction of the engines was begun at the Soho Works; in Philadelphia, work was started on the reservoirs, tunnel, and the elegant Greek revival enginehouse in Centre Square. After encountering many technical and financial snags, the engines were started in operation in December 1800 and January 1801.

COMPARISON OF EARLY AMERICAM STEAM ENGINES[23]

Engine	*Diameter of Cylinder (in inches)*
Philadelphia Waterworks (Schuylkill)	40
Schuyler Mine (estimated)	34½
Philadelphia Waterworks (Centre Square)	32
Hope Furnace	30
New York Waterworks (Colles)	18

When completed, these two engines were magnificent in appearance, if not in operation. The first engine, with a 40-inch cylinder, pumped Schuylkill River water from a low basin at the foot of Chestnut Street through a tunnel 6 feet in diameter and 3,144 feet long to the enginehouse at Centre Square. Here the second engine, with a 32-inch cylinder, pumped the water into a reservoir for ultimate distribution throughout the city.

With the engines in operation, mechanical failures were almost an everyday occurrence. The chief offenders appear to have been the boilers, constructed of white pine plank, which leaked badly almost immediately and were soon beset by rot. The Centre Square boiler was repaired and kept in use until 1814. The one at the Schuylkill basin was replaced in 1803 by a boiler "composed of 70 plates of iron, cast with flanches, and bolted together."[24] In both engines, the lever beams, the arms and shafts of the flywheels, the bearings, and other parts, being made of wood, were subject to rot and eventually had to be replaced. Latrobe became dissatisfied with the operation of the air pump of the Schuylkill engine and replaced it before 1804.

Both of these Roosevelt engines were replaced in 1816 by new ones built by Oliver Evans. During these years the waterworks had cost the city $657,398.[25] Many years later the chief engineer of the Philadelphia water department sought to justify this expenditure in terms of its "many indirect advantages." He traced "the pre-eminence which this city has always enjoyed in the construction of machinery" back to the experience gained from these early engines and declared that "there is scarcely a water-works in the construction of which Philadelphia mechanics have not, and are still contributing more largely than any others."[26] Although the engines were built at Roosevelt's Soho Works in New Jersey, Philadelphia reaped the greatest harvest of experience.

Not all of the English engineers who came to America during these years worked for Nicholas Roosevelt. Marc Isambard Brunel, who later won fame for his tunnel under the Thames at London, was in the United States from September 1793 to January 1799. He spent much of this time in New York and consulted with Robert R. Livingston on the subject of steam in 1797.[27] A Mr. Tongue, said to be the son of the Duke of Bridgewater's chief engineer, was sought out by William Thornton to give information on steamboat experiments in England.[28] Sometime after 1807 a foreman from the Boulton and Watt Works, brought to this country by Robert Fulton, was operating a steamboat on Lake Champlain.

Some of these emigrant engineers found service where their talents were used and appreciated. Latrobe boasted in 1812 that he had found "an English regularly bred Engine builder" who was sober, quiet, and exact in his work. He was made foreman of Latrobe's new works and proved himself a "good smith, and excellent Turner in Brass, Iron, & wood" and, best of all, he was "master of all the English improvements of the last 4 or 5 years."[29] As long as English practice remained the pattern for American development, those engineers who moved westward across the Atlantic and cut themselves off from a rapidly changing art soon found their knowledge becoming obsolete and their counsel outdated. There was no reason why they could not follow British developments at long range, but apparently some did not. Expatriated British engineers seemed to retain for years a vested interest in the state of the art as it had been when they emigrated to America.

Another well-qualified English engineer was a Mr. Coxen who in 1810 appeared in Washington, D.C. and sought out William Thornton. From his position as head of the federal patent office, Thornton served as a clearinghouse for much of the gossip and activity current among American engine builders. The English engineer presented letters of introduction and the two men immediately fell into happy conversation on technical matters. The Patent Commissioner pointed out models submitted for various patents, and Coxen demonstrated wherein they were soundly conceived and where they erred. He "made distinctions very much to my satisfaction," the Commissioner later wrote to a friend. Thornton was impressed with the fact that "Coxen . . . has made many Steam Engines on the best & most improved principles of Watts & Boulton's," and while in America would "engage to make them equal to any that can be imported." Mr. Coxen, he concluded, "has been engaged in some of the grandest works in England, and I can with truth say he is the only Engineer who has ever given me complete satisfaction."[30] Not only was Coxen well informed but, when he arrived in this country, his information was much more recent than that of Smallman, who had been away from the Boulton and Watt enterprise for at least fourteen years. After talking with Coxen, Thornton decided that Smallman's engines were "too much on the old plans," and the one he had built for the Washington Navy Yard in 1810 did not even use the parallel motion of the beam, a masterpiece Watt had developed in 1785-86.[31]

The appearance of these English engineers in America led Thomas Cooper, who was himself of English birth and training but now professor of chemistry at Dickinson College in Pennsylvania, to hope that "the time is now come, when a *civil-engineer* will be in general demand, and will meet with the success in point of fortune, and the respect in point of situation in society, which the able men of this description are sure to find in Great Britain." From his position as editor of the severely uncompromising *Emporium of the Arts and Sciences,* Cooper was unceasing in his attempts to act as broker between science and manufactures. He saw the "civil-engineer" as the link between them and laid down what he considered to be the requirements for that profession: knowledge of mathematics, chemistry, drawing, and practical experience.[32] Probably no engineer trained in America, with the possible exception of a few West Point graduates, could meet these high standards, but Benjamin Henry Latrobe came very close.

When Latrobe submitted his plan for the Philadelphia Waterworks in 1799, he immediately formed what was to become his normal pattern of relationship to engine making: that of consulting engineer. It is quite possible that this was, at least in part, dictated by his ignorance of the minutia of engine fabrication. While still a young man in England he had served an apprenticeship under John Smeaton, and his father had been an acquaintance of Matthew Boulton, but there is no evidence that Latrobe ever actually concerned himself with engine building. Another contributing factor was his temperament, which drew him more to the role of planner and adviser than to that of careful businessman and manager. Yet another element, less easy to document but nonetheless compelling, was the fact that his knowledge of his own superior training and abilities fitted him for a tutorial role in America. He was forced to grope toward a new profession—that of consulting engineer—because he felt himself overtrained for any other.

Latrobe's position as chief engineer of the Philadelphia Waterworks brought him to public attention and established him as an arbiter in matters relating to steam. In 1804 he answered a question sent to the American Philosophical Society by the Society of Rotterdam: "Whether any, and what improvements have been made in the construction of Steam-Engines in America?"[33] His answer was at best undistinguished and at worst marred by conservatism and prejudice. The report professed to see no improvements other than those made by Roosevelt and Latrobe. Furthermore, he denied the utility of steamboats and earned the undying antagonism of Oliver Evans by limiting notice of him to one deprecatory sentence. Although Evans at this time was a projector rather than a successful builder of engines, Latrobe's poor opinion not only was premature but also persisted long after its error was obvious to others. The report, despite its inferior quality, further entrenched Latrobe as the nation's unofficial expert on matters of steam.

By the time this report was made, Latrobe had ended his formal connections with the waterworks and was acting as chief engineer of the Chesapeake and Delaware Canal and as surveyor of public buildings in Washington. He retained his interest in the steam-powered rolling mill which, with Roosevelt, he had appended to the waterworks in Philadelphia.[34] James Smallman had been sent from Soho to operate the mill and used the opportunity to set up in the engine-making business for himself. When in the future Latrobe received inquiries about the possible use and purchase of engines, he referred the questioners to Smallman and Roosevelt for particulars, although he was always ready to offer advice and to write letters of introduction. This procedure was followed even after Smallman no longer had any formal ties with Latrobe and was manufacturing engines on his own. Latrobe expressed the hope that "my opinion is impartial," and claimed to have "no interest to bias" him. He recommended Smallman to all comers because he considered him "the best Engine maker in America."[35]

Despite occasional temptations, Latrobe refused to enter the business world as a maker of steam engines. To one inquiry he answered: "You have . . . entirely mistaken my profession if you suppose I have any thing to do with steam Engines further than to *use* them when I want them."[36] He explained that "when I want one for any public purpose, I contract for it with Smallman or Large, or Oliver Evans, or some other man whose trade is the making of Engines, & I have nothing to do then but to receive them when finished. . . . With making them I never have had & never intend to have any thing to do."[37]

Although he avoided the temptation to use engines for himself or to make them, he frequently bought them for public purpose, as with the two engines for the Philadelphia Waterworks. Latrobe had Smallman build an engine for the Navy Yard in Washington; the contract for constructing the waterworks at New Orleans was also received by Latrobe.

In 1807 a correspondent of Oliver Evans suggested that he and Robert Fulton undertake the New Orleans work, adding that "it is an important job, and he who secures it, will do a good work for his fellow Citizens, and will make his fortune."[38] Latrobe got the contract, however, and, far from making his fortune, it probably cost him one and, more tragically, took the life of one of his sons, who was sent to New Orleans to prosecute the work on the spot. Latrobe remained in Washington, D.C., coordinating the work, raising money (mainly from friends in New York and Baltimore) to finance the project, and arranging with Smallman in Philadelphia to build the necessary engine.

He soon became dissatisfied with Smallman—probably because of the latter's high prices—and cast about for some other means of obtaining the engine.[39] He toyed with the idea of buying a used engine—two of which were offered him at half the cost of new ones[40]—but finally he decided to build an engine himself. Contrary to his own firm intention and better judgment, he constructed a shop in Washington to perform the work.

With the outbreak of the War of 1812, the British blockade made transportation by sea to New Orleans impossible and Latrobe moved to Pittsburgh. He hoped to complete the engine there and ship it down the Mississippi River to his waiting son, but unfortunate business dealings with Robert Fulton thwarted these plans.[41] Latrobe returned to Washington having abandoned the idea of building his own engine. By this time he was desperate enough to think of negotiating with Evans for one of his high-pressure engines, and claimed that Evans had suggested such a plan.[42]

"With Smallman I believe nothing is to be done," he wrote in August 1815,

"& I do not intend to try Evans yet."[43] Two years previously he had met Robert McQueen, "the principal Engine builder at New York," and now in 1815 he settled on him to build the long-delayed New Orleans engine.[44] McQueen offered to travel to Washington to make personal arrangements for the construction, but most of the details were worked out through correspondence.

As was his habit, Latrobe made detailed drawings of the engine he wanted built. He explained to McQueen, however, that the drawings were "of no importance whatever to the proper execution of the work, excepting in as far as they explain the general principles on which I have hitherto proceeded." He confessed that McQueen had "more practical experience than I have" and hoped that he might "suggest many better methods of execution than my drawings describe." Together the two men discussed the proper proportions between air pump and cylinder and evaluated the various types of valves which might be chosen for use on the engine.[45] Although Latrobe later claimed that "from his experience I have derived benefit," these detailed discussions proved largely academic when he had to cancel this order and settle for a used engine, bought in New York through McQueen's agency.[46]

Throughout these discouraging and complicated attempts to find an engine, Latrobe's major role remained the same. He had taken the contract for the waterworks and attempted to subcontract the engine. In a better developed market for steam engines he might have merely placed an order for a standard size engine with a reputable firm and taken delivery in a reasonable length of time. As matters stood, however, two things, aside from the fortunes of war and the narrowness of his own financial base, prevented this simple arrangement. First, no one yet made standard size engines of predictable duty. Second, Latrobe's own conception of himself as innovator and professional engineer made any such direct dealings highly improbable. Latrobe used the word "innovation" to mean a change which, while reasonable in theory, had not yet been shown in practice to be an improvement. The proper relationship, therefore, between Latrobe and McQueen or Smallman, would be based on a clear division of labor. The engineer, drawing upon his knowledge of mathematics, chemistry, and drawing (to use Cooper's criteria), would suggest changes in engine design to the builder. He in turn would incorporate these and test them in practice. Presumably if the "innovation" proved to be a true "improvement," future engines would carry the same change.

The implications of such a relationship are many, though inherently difficult to document. From the standpoint of the enginemaker, while improvements were always sought in theory, the result of constant innovation in practice could only be subversive of standardization and routine and, therefore, of profits. The coolness which developed between Smallman and Latrobe can probably be traced in part to the fact that it was much easier for the enginemaker to deal directly with a purchaser ignorant of steam theory than through the agency of an engineer like Latrobe, who was constantly trying to innovate.

Another service performed by Latrobe fitted well with his position as leading steam engineer in the country. Like Thornton, in the patent office, he served as an intermediary between those whose inclinations or livelihoods led them to an interest in steam. On some occasions he brought together engine builders and iron founders who had the necessary facilities for casting parts. He gave free and detailed opinions on new developments in steam technology and on the application of steam power to manufactures. When the Massachusetts inventor Jacob Perkins sought information, Latrobe gave an answer and then, so that he could pursue the inquiry, provided him with letters of introduction to Robert Fulton and David Prentice, an iron founder.[47] In correspondence with Thomas Cooper, he exchanged bibliographic information on steam engines and introduced a Mr. Dowers, with whom Cooper later conducted experiments on what were identified as "vapor engines."[48]

Latrobe was atypical of the English engineers who arrived in America at this time only to the extent that he was better trained, more active, and more ambitious. These men had the potential, more or less developed, of making unique contributions to the establishment of steam technology in this country. Their frequent immigration kept America abreast of English developments, and they trained a generation of American mechanics to build and operate engines. When they coupled ability with learning and culture, as in the case of Latrobe, they raised the status of technology and made it a subject of serious consideration. In steam, as well as in other technologies, they brought the Industrial Revolution to America.

Notes

[1] "Report on Manufactures," *Alexander Hamilton's Papers on Public Credit, Commerce, and Finance,* edited by Samuel McKee (New York, 1957), page 208.

[2] Ibid., page 207.

[3] Loree, *First Steam Engine,* page 9; Nicholas Roosevelt to John H. B. Latrobe, 29 March 1828, Robert Fulton papers, NYHS, New York, New York.

[4] B. H. Latrobe, *American Copper Mines* (no place, [1800]), pages 5, 6; Loree, *First Steam Engine,* page 10.

[5] Matthew Boulton offered to buy the entire production of the Schuyler mine, and also to build the "adventurers" an engine to keep it properly drained. Matthew Boulton to Elias Boddinott, 8 June 1799, Record Group 104, Records of the Bureau of the Mint, Boulton Correspondence, National Archives, Washington, D.C.

[6] Allan Nevins, *Abram S. Hewitt: With Some Account of Peter Cooper* (New York, 1935), pages 3-4, 6.

[7] *Spectator* (New York), 4 October 1797, in Gottesman, *Arts and Crafts in New York,* page 261; Nicholas Roosevelt to John H. B. Latrobe, 29 March 1828, Fulton papers, NYHS; John H. Morrison, *History of American Steam Navigation* (New York, 1958), page 42; Turnbull, *Stevens,* page 129.

[8] Nicholas Roosevelt to John H. B. Latrobe, 29 March 1828, Fulton papers, NYHS.

[9] B. H. Latrobe to Mrs. Roosevelt, 31 December 1810, B. H. Latrobe MS. Coll., Baltimore, Maryland.

[10] Nicholas Roosevelt to Aaron Ogden, 7 November 1814, Fulton papers, NYHS.

[11] [B. Henry Latrobe], *An Answer to the Joint Committee of the Select*

and Common Councils of Philadelphia, on the Subject of a Plan for Supplying the City with Water, &c. (no place, 2 March 1799), page 3; *Proposals for Establishing a Company, For the purpose of employing the surplus power of the STEAM-ENGINE Erected near the river Schuylkill: Under the Title and Firm of the PHILADELPHIA ROLLING COMPANY* (no place, [about 1800]), page 3. The assertion that Smallman was associated with Boulton and Watt has never been questioned although it rests only on the statements of Latrobe and Roosevelt.

[12] Nicholas Roosevelt to John H. B. Latrobe, 29 March 1828, Fulton papers, NYHS.

[13] Deposition of Nicholas Roosevelt, 14 January 1815, Fulton papers, NYHS.

[14] R. R. Livingston to Nicholas Roosevelt, 8 December 1797, Fulton papers, NYHS.

[15] Deposition of Nicholas Roosevelt, 14 January 1815, Fulton papers, NYHS.

[16] Copy of testimony of James Smallman, Evans papers, Franklin Institute.

[17] Nicholas Roosevelt to Aaron Ogden, 7 November 1814, Fulton papers, NYHS.

[18] R. R. Livingston to Nicholas Roosevelt, 17 August 1798, Fulton papers, NYHS.

[19] R. R. Livingston to James Watt, 4 November 1798, quoted in H. W. Dickinson, *Robert Fulton, Engineer and Artist: His Life and Works* (London, 1913), page 135.

[20] James Mease, *The Picture of Philadelphia, Giving an Account of its Origin, Increase and Improvement in Arts, Sciences, Manufactures . . .* (Philadelphia, 1811), page 149.

[21] *Pennsylvania Gazette,* 14 March 1799.

[22] [B. Henry Latrobe], *View of the Practicability and Means of Supplying the City of Philadelphia with Wholesome Water. In a Letter to John Miller, Esquire, from B. Henry Latrobe, Engineer. December 29th. 1798* (Philadelphia, 1799), pages 7-8, 20. Interestingly enough, this was the same argument used later by Oliver Evans to justify the use of the high-pressure rather than the Watt engine.

[23] For the size of the Philadelphia engines see "History of the Steam Engine," *Journal of the Franklin Institute,* volume 102 (1876), page 256 and Latrobe, "First Report," *Transactions of the American Philosophical Society,* volume 6, part 1 (1804), page 95. For the Hope Furnace, Cutler, *Journals,* volume 1, page 206. For the Colles engine [Bangs], "1776," *Historical Magazine,* volume 4 (1868), page 306. It should be remembered that the Philadelphia engines were double acting, thereby giving, theoretically, double the power.

[24] Frederick Graff, "Notice of the Earliest Steam Engines Used in the United States," *Journal of the Franklin Institute,* volume 55 (1853), page 207; Latrobe, "First Report," *Transactions of the American Philosophical Society,* volume 6, part 1 (1804), pages 95-96; "History of the Steam Engine," *Journal of the Franklin Institute,* volume 102 (1876), page 264.

[25] *Aurora* (Philadelphia), 18 May 1816. Tanner estimates that during this same time water rents brought in $105,351. *Description of the Canals,* page 105.

[26] Department for Supplying the City with Water, *Annual Report of the Chief Engineer of the Water Department of the City of Philadelphia For the Year 1875* (Philadelphia, 1876), page 18.

[27] David Read, *Nathan Read: His Invention of the Multi-Tubular Boiler*

and Portable High-Pressure Engine, and Discovery of the True Mode of Applying Steam-Power to Navigation and Railways (New York, 1870), page 136.

[28]William Thornton, "Answers . . . to numerous Interrogations," no date, Thornton papers, volume 16, LC; memo, unsigned, no date, Thornton papers, volume 4, LC.

[29]This engineer, William Robinson, is described in B. H. Latrobe to Henry Latrobe, 9 August 1812, and same to same, 6 September 1812, B. H. Latrobe MS. Coll., MHS.

[30]William Thornton to John Gavin, 14 December 1810, Thornton papers, volume 3, L.C.; William Thornton to James Van Ingen, 14 December 1810, Thornton papers, volume 3, LC. Coxen, the next year, built a steamboat for service on the Delaware River. Turnbull, *Stevens,* page 295.

[31]William Thornton to James Van Ingen, 14 December 1810, Thornton papers, volume 3, LC.

[32][Thomas Cooper], "An Account of the Steam Engine," *Emporium of the Arts and Sciences,* volume 2, new series (April 1814), page 385.

[33]Latrobe, "First Report," *Transactions of the American Philosophical Society,* volume 6, part 1 (1804).

[34]See Fritz Redlich, "The Philadelphia Water Works in Relation to the Industrial Revolution in the United States," *Pennsylvania Magazine of History and Biography,* volume 69 (July 1945), pages 243-256.

[35]B. H. Latrobe to Andrew Hazelhurst, 19 October 1810, B. H. Latrobe MS. Coll., MHS.

[36]B. H. Latrobe to Robert Smith, 4 July 1811, B. H. Latrobe MS. Coll., MHS.

[37]B. H. Latrobe to Robert Smith, 5 August 1811, B. H. Latrobe MS. Coll., MHS.

[38]Richard Claiborne to Oliver Evans, 24 August 1807, Evans papers, Franklin Institute.

[39]B. H. Latrobe to Robert Fulton, 12 August 1811, B. H. Latrobe MS. Coll., MHS.

[40]B. H. Latrobe to Daniel Large, 10 August 1811, B. H. Latrobe MS. Coll., MHS; B. H. Latrobe to James Smallman, 10 August 1811, B. H. Latrobe MS. Coll., MHS.

[41]B. H. Latrobe to Robert Fulton, 5 July 1812, B. H. Latrobe MS. Coll., MHS.

[42]B. H. Latrobe to Frederick Graff, 12 August 1815, B. H. Latrobe MS. Coll., MHS.

[43]B. H. Latrobe to Jacob Mark, 12 August 1815, B. H. Latrobe MS. Coll., MHS.

[44]B. H. Latrobe to Henry Latrobe, 18 September 1815, B. H. Latrobe MS. Coll., MHS; B. H. Latrobe to Frederick Graff, 27 August 1815, B. H. Latrobe MS. Coll., MHS.

[45]B. H. Latrobe to Robert McQueen, 17 October 1815, B. H. Latrobe MS. Coll., MHS; same to same, 5 October 1815, B. H. Latrobe MS. Coll., MHS.

[46]B. H. Latrobe to Henry Latrobe, 8 November 1815, B. H. Latrobe MS. Coll., MHS; B. H. Latrobe to Samuel Ellis, 3 May 1816, B. H. Latrobe MS. Coll., MHS; B. H. Latrobe to Jacob Mark, 3 May 1816, B. H. Latrobe MS. Coll., MHS.

[47]B. H. Latrobe to Robert Fulton, 16 March 1813, B. H. Latrobe MS. Coll., MHS; B. H. Latrobe to Jacob Perkins, 24 March 1813, B. H. Latrobe MS. Coll., MHS.

[48]B. H. Latrobe to Thomas Cooper, 8 February 1814, B. H. Latrobe MS. Coll., MHS; same to same, 1 March 1814, B. H. Latrobe MS. Coll., MHS; [Cooper], "Account of the Steam Engine," *Emporium,* volume 2, new series (1814), page 376.

Chapter 4

Early Engine Makers

[I have] . . . in 3 years last, done
more to acquire permanent property,
by renouncing such [inventive] persuits,
and following regular
business, than in 30 years before.
OLIVER EVANS (1809)[1]

Of the more than a dozen inventors in the United States who had built steam engines before 1800, none had engaged in their manufacture for general sale. These men built engines for application to some particular purpose, usually to operate a steamboat. They built their own engines because no others were available. Interest in steam waned when the anticipated experiment failed, and the experience gained was not applied to building engines for those who might have had some use for them. Men such as John Fitch and William Longstreet were inventors first and engine builders secondarily and only out of necessity.

Engine shops that were building for the general trade–steamboat or stationary–were operating in Philadelphia and New York by 1812. Both cities were centers of mechanical arts and had been centers of early steamboat activities; they also had steam-powered municipal waterworks. These factors contributed to making them the first centers of engine building in the country.

Philadelphia led early in the manufacture of stationary steam engines. Its congregation of scientific, financial, and governmental leaders during the 1790s had made this city the natural focus of steamboat speculation. It was the center of the mechanical arts in America and the site of Fitch's most active work. The two great waterworks engines that Benjamin Henry Latrobe had erected characterized the city's early interest in steam. By 1808 the enthusiastic claim was made that "there is no part of the world in proportion to its population, where a greater number of ingenious mechanicks may be found than in Philadelphia and its immediate vicinity.

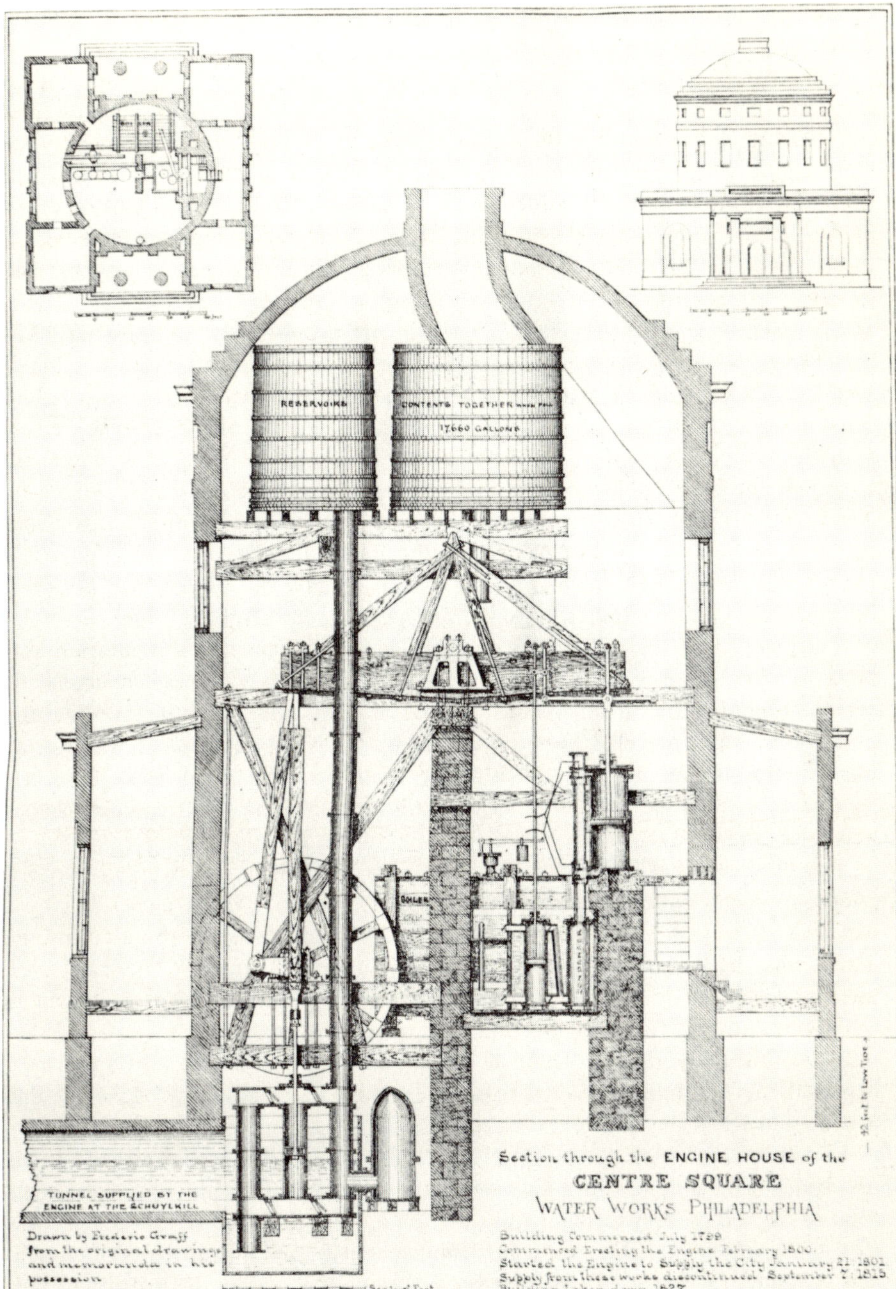

Engine built by Nicholas Roosevelt for the Centre Square Waterworks, Philadelphia, 1799. From *Journal of the Franklin Institute*, volume 102 (October 1876), plate 1.

Steam engines with all their various improvements are built and applied, beneficially, to the most useful purposes."[2]

When Nicholas Roosevelt built the waterworks engines in 1799-1800, he had sent James Smallman, his most trusted and experienced workman, to supervise their erection and early operation. Smallman forsook the Soho Works in New Jersey for Philadelphia where he remained in charge of the rolling and slitting mill which Roosevelt–with Latrobe and other partners–had built to use the surplus power of the Schuylkill (lower) engine.[3] By 1804, while still connected with the Philadelphia Rolling Mill, he laid plans for building stationary steam engines of all sizes and varieties for general sale.[4]

The decision to make engines for sale was not taken lightly. In February 1804 the partners were quoting a price of $7,450 for an engine with a 24-inch cylinder, double acting, complete with beam, flywheel, sun-and-planet gears, shaft, and cast-iron boiler, and "*put up* completely, but independently of the *house & frames*, & of the setting of the boiler." [5] But uncertainty continued. "The risk and expense attending the construction of one or two Engines is too considerable," Latrobe confessed, "to permit it to be a matter of anxiety to myself & my partners to invite that sort of business, by low charges."[6] Smallman, nevertheless, did build a few engines during the next several years, but one built for a mine in Pottsgrove, Pennsylvania, performed poorly and did not help his reputation.[7]

Smallman's most famous engine, and one which did enhance his reputation, was built for the Navy Yard in Washington, D.C. Latrobe, who was in charge of construction at the Yard on Anacostia Creek, informed Smallman in 1809 that the purchase of an engine had been authorized by the Navy Department. The engine was delivered ten months later, in June 1810, and Latrobe found it executed in a manner to exceed expectations. Five of Smallman's workmen (including James Smallman, Jr.) were sent from Philadelphia to help put up the engine.[8]

The original Navy Yard engine had a 21-inch cylinder and worked with 12 pounds of steam on the safety valve, supplied by a boiler 11 feet 8 inches long by 6 feet wide. It was rated at about 14 horsepower. Latrobe had incorporated into its design what he called the "double American airpump," probably the same type used by Roosevelt in his steamboat of 1798 and by Latrobe in the Philadelphia waterworks engines. William Thornton saw the engine in December 1810 and said that Smallman's work was "good & substantial, yet he has not introduced as I expected the last Improvements, particularly the parallel motion."[9] Like the one proposed in 1804, this engine was equipped with the sun-and-planet gears.

For all its old-fashion design, the Navy Yard engine worked well. Latrobe had his mechanics make innumerable adjustments and minor changes until, after spending a total of $8,500, he considered it to be in "the most perfect order." This engine was to drive a scrap-metal stamper, forge hammer, grind stones, bellows, a block mill, and sawmill;[10] it was used until the Navy Yard was burned by the British in 1814. The engine was then rebuilt and used until 1830, when its 14 horsepower was inadequate for the increased workload. It was replaced by a 30-horsepower engine of more modern construction but poorly made and in four years was replaced by a 40-horsepower, high-pressure engine.[11]

Centre Square Waterworks Enginehouse, 1799.
From John F. Watson, *Annals of Philadelphia, and Pennsylvania, in the Olden Time . . .*(Philadelphia, 1881), volume 2, page 457.

Latrobe continued to recommend Smallman as an enginemaker, and even Thornton suggested his name to friends, although cautioning them that Smallman would "require very particular directions in every part of the work."[12] In 1811 the cost of an engine, "on the Wharf at Philadelphia, exclusive of the fireplace or furnace of the boiler which any Smith can make," was six thousand dollars.[13] When Latrobe was planning his New Orleans waterworks engine in 1811, he contracted with Smallman for the work but later had to cancel out, buying the parts which had already been made. Smallman meanwhile was trying to consolidate his position as Philadelphia's leading builder of steam engines of the Boulton and Watt type, but he had a ready and able competitor in Daniel Large.

Large had come to the United States about 1807 and settled in Philadelphia. In 1808 he built and installed a large (10-horsepower), low-pressure engine to grind paint at the white-lead works of Wetherill and Brother.[14] Latrobe heard that this engine was "the best hitherto in America" and, on a trip to Philadelphia, went with Smallman to see it in operation. Upon arriving at the lead works, Latrobe was given free access to the enginehouse, but Smallman was excluded because "Mr. Large himself had some idea of going into the business of building Engines," and Large, who was a foreigner and had no patent, did not want to give away the nature of his improvement. The improvement was not made explicit, but was probably the parallel motion, which Smallman still had not adopted.[15]

Latrobe pointed out that Large, with his water-driven lathes and other equipment in Germantown, and Smallman, with his shop and tools in Philadelphia, would benefit by a joining of forces to manufacture low-pressure engines. Smallman proceeded to honor the suggestion in fact but not in spirit. He became Large's partner by loaning him a much-needed ten thousand dollars. Within a short time, however, he sued to recover the money, thus forcing Large into bankruptcy. Latrobe was embarrased by this turn of events and offered Large a salary of "1,500 a yer plus free rent" to come to Washington and manage his new engine shop. This offer was rejected.[16]

Large continued to operate in Philadelphia, making both stationary and steamboat engines. In 1812 he offered to enter into partnership with William Thornton, but nothing came of this proposal.[17] In 1816 Large built a steamboat under an arrangement with David Prentice.[18] Three years later he went into business with Samuel Humphreys and William Jones to build another steamboat, a venture which not only helped illuminate his character but also gave some indication of the problems of capital and market faced by early builders.

Large suggested the scheme and was to build the engine. A first draft of an agreement to that effect was drawn up, showing that Humphreys was to build the hull, Large the engine, and Jones, a merchant, was to pay for wages and material. Large objected because he thought he could not keep the cost of the new engine separate from that of the several engines already under construction in his shop. Another contract was drafted which bound Large to provide the engine and machinery and to install them on board the boat for $8,500, which was to be his share of the partnership. Again Large demurred, apparently suspicious of the whole venture. Finally the other two men decided to proceed without a contract and to buy the engine from Large when the boat was ready. Eventually, however, they sold the hull without any engine aboard.[19]

By 1838 at least eight different engines were operating in Philadelphia, built either by Daniel Large, C. Large (possibly Daniel's son), or the two in partnership. One was for a steamboat, the others were for manufacturing purposes. In 1836 C. Large established the Soho Works in Philadelphia for the manufacture of high-pressure steam engines.

Smallman built engines in Philadelphia until 1822.[20] Latrobe continued to refer business to him for awhile but by 1813 was strongly hinting that although "the principles of Bolton & Watt, on which you construct your Engines have not been, I think, improved," the old sun-and-planet gearing still used by Smallman was becoming very much dated.[21] The ultimate eclipse of Smallman has never been explained, but his technical conservatism, plus the rise in popularity of the high-pressure engine, no doubt contributed.

Latrobe consistently and at times even passionately recommended the engines of Smallman over those of Oliver Evans, but in the end, Evans was more successful. Evans, who came to be called "The Watt of America," was born in Delaware in 1755. After apprenticing as a wheelwright, he clerked in a general store. In 1782 he purchased a merchant flour mill in which to construct and operate his most famous invention, the automatic miller. This remarkable series of improvements routed and processed grain into flour entirely by means of machinery moved by waterpower. As this device became accepted, Evans moved to Philadelphia where, about 1793, he opened a millers' supply house and millstone factory.

As a young man, Evans amused himself and his friends by shooting corks from heated, water-filled gun barrels. He later claimed that this demonstration suggested to him the expansive power of steam and set him to speculating on its usefulness. When he subsequently found a description of an atmospheric steam engine, he professed astonishment "to find that the steam, was not applied as the prime mover, but only as the means to form a vacuum, to apply the weight of the atmosphere. This he conceived to be a great error."[22] He experimented with a rotary engine in 1784, but abandoned this line and concentrated on the development of his famous Columbian high-pressure steam engine.

The old Newcomen engine had used condensed steam to produce a vacuum in one end of a cylinder which caused the atmosphere to push the piston down. Watt had modified this action by attaching a separate condenser so that the temperature of the cylinder would not be lowered at each condensation. The piston, however, was still "pulled" by the vacuum at each end of the cylinder. In Evans' Columbian engine, steam that was introduced into the cylinder under high pressure was allowed to expand and push the piston to the opposite end. As the piston moved from one end to the other, spent steam was expelled. Thus Evans was able to dispense with the condenser and to increase the power of his engines merely by increasing the pressure of the steam from the boilers.

Figure 3 shows the essential characteristic parts of Evans' noncondensing, high-pressure, Columbian steam engine. Steam was raised in the boiler (*A*) to 100-200 psi, and routed by the ingenious rotary valve (*F*), through a pipe (*H'*), into the cylinder (*B*) above the piston (*G*). As the high pressure of the steam above drove the piston downward, the spent steam below the piston was forced through a pipe (*HH*), and the rotary valve (*F*), to escape into the air. When the piston (*G*) reached the bottom of the cylinder, the rotary valve was turned clockwise 90 degrees, allowing steam from the boiler to enter the bottom of the cylinder through pipe (*HH*). The spent steam above the piston was now able to exhaust through the pipe (*H'*), and the rotary valve. The grasshopper beam (*C*) rested on its fulcrum (*D*) and, when raised and lowered by the piston rod (*J*), turned the flywheel (*E*) by means of a crank and connecting rod (*K*).[23]

Evans considered "that this invention will, when applied to all the useful purposes, yield ten times more benefit to my country than my improvements on the manufacture of flour."[24] From 1801, when he began to build his first full-scale engine, until his death in 1819, the development and manufacture of this engine absorbed his major interest and activity. Expensive and time-consuming patent-infringement suits, growing out of his mill improvements, continued throughout the period. His interest in these, however, was prompted by his desire for the just rewards of invention rather than any interest in improving or manufacturing.

Plaster of paris, a staple commodity in Evans' Philadelphia store by 1799, was being used increasingly both for stucco work and as a farm manure. It was exceedingly difficult to grind, even with the screw mill which he had invented in

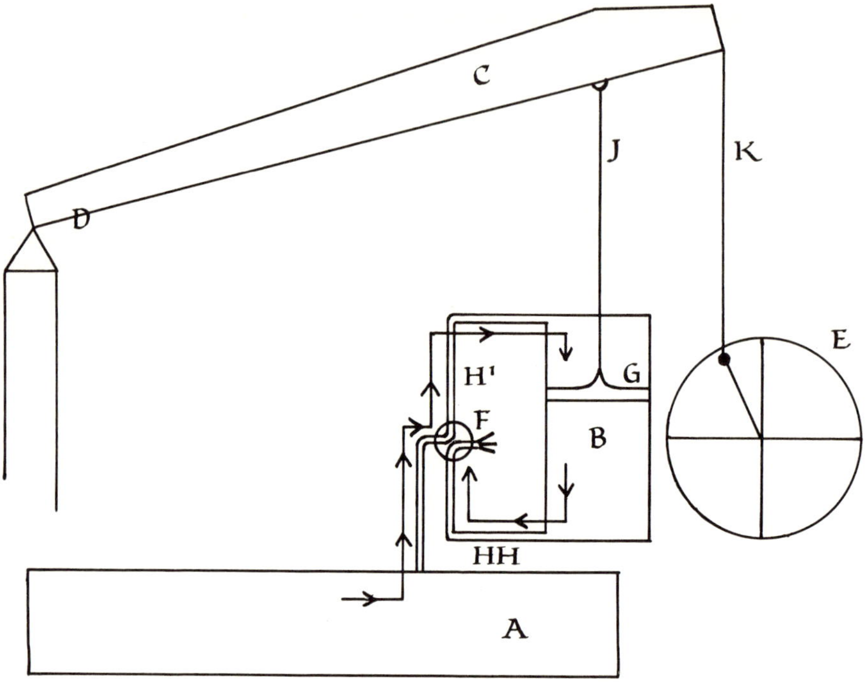

A Boiler
B Cylinder
C Grasshopper beam
D Fulcrum of beam
E Flywheel
F Rotary valve
G Piston
H and HH Pipes
J Piston rod
K Driving rod

Figure 3 Evans noncondensing, high-pressure Columbian steam engine, 1804

1795-96. By 1801 he had proceeded so far with his plans for a high-pressure engine that he was able to begin building one to grind his plaster. This engine was completed in the winter of 1802 and, as he said, "excited considerable attention and curiosity."[25]

That same year Evans began building his first engine for sale. A Mr. Louis Vancourt came to Philadelphia to order an engine for a projected steamboat to run between Natchez and New Orleans. Before the boat could be placed in operation it was grounded by low water, and the owners sold the engine to William Donaldson who used it to drive a sawmill. This mill was burned down a short time after, but

Donaldson bought another engine in 1806 and renewed successful operations.[26] Meanwhile the small (6-inch cylinder) engine used to grind plaster in Philadelphia was advertised to the public as able to serve "all the various purposes of machinery."[27] Interested persons were invited to view and judge its merits.

In the fall of 1803 Evans built a smith shop adjoining his millstone factory at Ninth and Market Streets, perhaps in anticipation of more engine work.[28] The next spring he announced that he was ready to construct engines. He put this operation under the charge of Charles Taylor, an experienced English engineer, who had lived in Philadelphia since 1797.[29] At this time Evans emphasized the small size and consequent cheapness of his engines and the fact that since the steam was not condensed, it could be used in other manufacturing processes which required heated water.

Orders for engines during 1805 were slow, and Evans took the opportunity to conduct an experiment that he had long wanted to attempt. This was the building of his steam-driven amphibious dredge, the *Orukter Amphibolos.* He had tried as early as 1786 to interest capitalists and state legislatures in the problem of steam carriages for use on common roads, particularly the new turnpikes. Having been consistently thwarted in these attempts, he contracted to dredge the wharf area of Philadelphia. The required mechanism had to navigate, to be sure, but the amphibious aspect of the *Orukter* was pure publicity. Before starting to dredge, it was driven around Centre Square, where the Latrobe-Roosevelt engine was still operating. The curious who came to see this new wonder were charged twenty-five cents for the privilege.[30] Evans pledged that half of the funds thus collected would be applied to defray the expense of new and useful inventions.

Another advertising and fund-raising venture of 1805 was the publication of Evans' major writing on steam, *The Abortion of the Young Steam Engineer's Guide.* Unlike his previous book, *The Young Mill-Wright and Miller's Guide,* which went through at least fifteen editions between 1795 and 1860, *The Abortion* was never reprinted, although in 1821 an American, Isaac Doolittle, published a French edition. The odd title evolved from Evans' having rushed the book to press before it was finished to his own satisfaction: a move dictated by his urgent need of money. This volume was the first guide of its kind written by an American and no doubt found a use in many engine shops.

As Evans' steam engine business increased, he began in 1806 to construct new shops in Philadelphia, and by the next spring the Mars Works was ready. It consisted of an iron foundry and shops for steam engineers, mouldmakers, and blacksmiths. In addition, the millstone factory was moved from Market Street to the Mars Works, along with the steam mill for grinding plaster and turning heavy cast and wrought iron.[31] The diversified products of the Mars Works included nearly all types of general iron work from sugar kettles to cast-iron cogwheels. These last came "from [a] mould which he has made for the purpose. They come," Evans advertised, "cheaper and will last longer than those made of wood."[32] Within a year the establishment was turning out patented steam engines, cogwheels, "from the smallest used in machinery for cotton, flax, and wool spinning, to the largest used for mills," sugar-mill rollers "neatly turned," screw mills, anvils, and similar items.[33]

Oliver Evans (1755-1819), inventor and engine builder.

A dramatic occurrence in May 1809 highlighted that year at the Mars Works. Evans had become unusually despondent over his position as inventor and self-styled technical tutor to the nation. A law suit verdict against him led to his comparing the rewards of inventing and manufacturing. "I was then in my 54th year," he wrote, "and had in 3 years last, done more to acquire permanent property, by renouncing such persuits [as invention], and following regular business, than in 30 years before." Calling his family about him, Evans explained that in order to remove from his sons the temptation to follow him as an inventor, he proposed to burn his papers. "They all approved," he recorded, "and I laid them on the fire." He hoped that now the way would be clear for him to "persue regular business the remainder of my life."[34]

The next few years were apparently ones of prosperity for Evans. Early in 1812 he reported that ten of his high-pressure engines had been built and put in operation since 1803. In addition, ten others were in various stages of completion in his shops.[35] It would seem that for the first nine years of his operations, Evans depended upon general mill ironwork to keep his shops busy. As time went on, however, the manufacture of steam engines became proportionately more important in his business, as is confirmed by the building of the Pittsburgh Steam Engine Company in that city in 1812. This factory was virtually a western branch of the Mars Works and indicates both the increasing prosperity of Evans and the growing importance of engine building in his plans.[36]

The Mars Works was destroyed by fire in April 1819. The value of the moulds alone that were burned was estimated at fifteen thousand dollars. Four days later, Oliver Evans died in New York City at the age of 64. Some years before, James Irwin Rush and Peter David Muhlenberg, sons-in-law of Evans, had entered the business. These two now took over the enterprise and, operating as Rush and Muhlenberg, continued as leading enginemakers until the death of Rush in 1845. The firm of Neall, Matthews, and Moore then carried on the business for another half century.

Even as Evans' enterprises were continued successfully after his death, a myth arose that he had died unhappy and poverty-stricken, denied by an ungrateful and parsimonious public. He himself fed this myth. "He that studies and writes on the improvements of the arts & sciences," he wrote bitterly in his personal copy of *The*

Oliver Evans's Columbian steam engine, 1813.
From *Niles' Weekly Register,* volume 3 (1813), Addenda, facing 1.

Abortion, "labours to benefit generations yet unborn, For it is improbable that his Contemparies will pay any attention to him." Again, one must distinguish between Evans the inventor and Evans the manufacturer. Perhaps this was the distinction Hezekiah Niles had in mind when he wrote that Evans had "been rendered almost misanthropic by what he, (as, we thought, erroneously), believed was the injustice and ingratitude of the public,"[37] with regard to his flour-mill patents. In fact his manufactures financed his patent fights and inventive efforts rather than the reverse.

Under the stimulation of Robert Fulton's activities, New York rose rapidly to challenge Philadelphia as a center of steam engine manufactures. The origins of the city had been commercial, and well into the nineteenth century trade rather than manufactures was New York's chief reason for existence. Indeed a guide book in 1827 admitted that "the island of Manhattan, on which the city lies, has no facilities for manufacturing pursuits, and the great natural advantages of New-Jersey and Connecticut with regard to manufactures, will for ever confirm the direction, which her capital and industry assumed from the first stage of her existence." Nevertheless, this same report significantly noted that of the total capital which was devoted to manufactures, the largest single amount was "invested in the manufacture of steam engines and castings of every description."[38]

The first engine in the city had been built by Christopher Colles for the municipal waterworks in 1775. By 1804 Nicholas Roosevelt had an engine–undoubtedly of his own construction–operating a sawmill on Rhinelander's Dock.[39] In that same year an engine imported from Boulton and Watt and owned by Alexander Hamilton's Manhattan Water Company was pumping water for the city.[40] Fulton too made inquiries to Boulton and Watt about the purchase of an engine as early as 1794. In 1803 he sent them rather exact specifications for an engine to power the contemplated steamboat *Clermont.* At this time he requested only certain parts of the engine, notably the cylinder, piston rod and piston, valves and their movements, air-pump piston and rod, and the condenser and its connections. "The other parts," he informed them, "can be made in New York, and as it will save the expense of transport, and they require a particular arrangement which must be done while I am present, I prefer having them done there."[41]

The ordered parts reached New York in several shipments during 1805 and 1806, and were stored until Fulton's return from England. The engine was then assembled and placed on board the *Clermont* for her successful trip in 1807. The Livingston-Fulton interests immediately began to build more steamboats which, with the help of Roosevelt's old workman Charles Stoudinger, were equipped with American engines. Two more boats were added in 1807 and then an additional twelve were built in New York before September 1816. Of these fourteen boats which followed the *Clermont,* five were built in 1813, the year Fulton built his large workshops and dry dock at Jersey City.[42]

Some of the work required for the Fulton steamboats was farmed out to other shops in New York City. Iron castings, for example, were purchased from Robert McQueen, a Scot millwright who operated an air furnace in New York. Partly because of the patronage and experience gained from his work for Fulton,

McQueen's steam activities and general casting business flourished. About 1811 he built engines for two boats owned by a company set up in Albany to rival the Fulton-Livingston monopoly. Sometime in 1813 he became acquainted with Benjamin Henry Latrobe, who contracted with him for the New Orleans waterworks engine and called McQueen "the principal Engine builder at New York."[43] In 1815 McQueen was "Engine Maker to the New York water works."[44] His shops were still in operation in 1820, by which time he had become an owner as well as a builder of steamboats. In that same year his Columbian Iron Foundry and Steam Manufactory represented an investment of $200,000. It employed seventy men and a dozen boys and used five steam engines to drive the lathes, boring machines, and other equipment. The product was described as "Steam Engines—Castings of all descriptions for Mills & Machinery—Stoves, &c Hollow ware and all Kinds of wr[ought] Iron Work, Screws—&c. and Steam Engine Boilers of wr[ought] Iron Cast Iron or Copper."[45]

While McQueen made many of Fulton's iron castings, his brass castings were obtained from James P. Allaire. Born in 1785, Allaire had learned the brass-casting trade from Louis Rhode, the former assayer at Roosevelt's copper mine. He worked for some time in New York in the brass foundry of a Mr. Ellsworth. When Ellsworth died in 1805 or 1806, Allaire took over the foundry, and it was here that he did his early work for Fulton. Like McQueen, Allaire prospered, partly because of Fulton's patronage. In 1813 he moved his foundry to the later famous Cherry Street location. Upon Fulton's death in 1815, Allaire joined with the ubiquitous Charles Stoudinger to operate the Jersey City shops. Stoudinger died the next year, and Allaire consolidated his enterprises by moving the Fulton equipment and business to Cherry Street. The Allaire Works continued in operation for many years and grew to be one of the largest in the country. In 1820 it had an investment of $100,000, employed sixty men and ten boys, and produced "Steam Engines & Boilers for Boats sugar & Other Mills, Iron & Brass Castings for all kinds of Machinery, as well as for Ship and domestic uses."[46]

Trained by Rhode and patronized by Fulton, Allaire's enterprise was firmly rooted in the mainstream of American steam technology. He passed this training on to men who became among the most important in the annals of American engineering. Among his many apprentices were John Roach (1815-1887), one of the country's leading builders of iron steamships; Louis De Coudres (1789-1872), for many years in charge of Allaire's brass foundry and one of the leading mechanics of New York for half a century; and Charles Haynes Haswell (1809-1907), first engineer-in-chief of the United States Navy.[47]

There were many similarities in the activities of early enginemakers—Smallman, Large, and Evans in Philadelphia and McQueen, Fulton, and Allaire in New York. None of them, for example, attained any great degree of specialization. Evans, McQueen, and Allaire did a considerable business in general metal work, especially in their early years. Evans, Smallman, Large, McQueen, and Allaire built (or planned to build) both stationary and steamboat engines. Fulton seems to have devoted himself exclusively to steamboats, although he was consulted occasionally about mill engines. Significantly, having no interest in doing general mill work and foundry business, Fulton found it worthwhile to send out for castings rather than invest in separate facilities of his own. Evans and Smallman, on the other hand, seldom involved themselves with marine engines. Smallman apparently did no more in that line after the Livingston-Roosevelt experiment in 1798 and of the twenty engines built or under construction at the Evans' Works in 1812, only four were for boats and one of these was never used for that purpose.

Most of these men included engines for public works in their planning. Smallman built the Washington Navy Yard engine and was involved with Latrobe in the New Orleans waterworks project. He had also been intimately involved with the building of the engines for the first Philadelphia waterworks. Evans too had been considered for the New Orleans job and had built the second Philadelphia waterworks project which replaced the old Centre Square facility. Latrobe had brought Large and McQueen into the New Orleans work, and the latter had been attached to the New York waterworks project. All told, public commissions were important in encouraging early steam engineering.

The relations between the various makers within each of the urban centers have been discussed, but communications between each of the groups were also important. Latrobe, in Washington, corresponded with all of the makers except Allaire. Incessant patent litigations kept the contacts fresh and made for subtly shifting alliances between all steam enthusiasts. Evans, having something of a monopoly on the high-pressure engine, was the most isolated. Even so, Fulton wrote him that "I wish to give one of your engines a fair tryal," and asked for the terms upon which one could be purchased.[48] Smallman, too, was approached by Fulton and claimed to have actually built an engine for him.[49]

A common problem shared by these men was one of financing the manufacture of the comparatively expensive steam engines. It should be remembered that for the first twenty years of their partnership, Boulton and Watt did not make their own engines (except for the valve gears). They had neither the trained workmen nor the necessary capital to engage in such an expensive and complicated enterprise. Their role during these years was to grant licenses, provide an erector, and act as consulting and coordinating engineers. The engines were actually built by various manufacturers, most notably John Wilkinson, the famous iron founder. The purchasers paid the founder or engineer directly, thus relieving Boulton and Watt from the necessity of tying up their limited funds. Even so, licenses, frequently granted to mining firms, sometimes had to be paid for in shares of stock rather than cash.

Allaire, McQueen, Roosevelt, and other early enginemakers, who were iron founders before they were engineers, already had most of the needed facilities and were able to do much of their own work. The early steamboat experimenters, like John Fitch, had to search out facilities for having large pieces (especially the cylinders) cast. At one point Fitch wrote that "I have since I began been Idle more than Six months waiting for the castings of a Cylinder and now am in doubts whether I can get it cast in america."[50] Similarly, Latrobe, working in Washington in 1813, had to make a personal trip to New York to arrange for the casting of a cylinder. Of his two usual foundrymen, one could not and one would not attempt the job.[51]

Such frustrations and uncertainties led most engine builders to do as many of the steps as possible. This in turn led to very large capital investments, not only in plant facilities but also in the engines themselves, before they were paid for. This placed them in a worse position than that occupied by the first American automobile manufacturers decades later, who bought all their parts on credit from special suppliers and then sold their cars for cash to dealers. Fulton was in a special position since he was able to draw upon the wealth of Robert R. Livingston; also the engines that were made were applied immediately to revenue-producing steamboats. Even so, Livingston was forced to sell his New Jersey lands to support the venture and was for a time faced with real money problems. The other two New York makers, McQueen and Allaire, had large facilities already available and were able to draw upon the profits of a general metal-working trade; furthermore, both were interested in steamboat companies.

In Philadelphia, Smallman got his start under the financial wing of Roosevelt and Latrobe. He charged six thousand dollars for an engine with a 21-inch cylinder, capable of powering a three-stone gristmill.[52] If only one or two such engines were being made at a time, a considerable amount of money was needed to carry on the business. He, like other builders, sought to finance an occasional engine by using it as an investment in a steamboat company, the cost to be paid from profits. In 1810 he proposed to William Thornton that they build such a boat.[53] As Thornton explained the proposal, Smallman "would be glad to undertake the works, and would take six shares at 100 Dolls. each, if I would allow him 1/20th of the Boat for superintending & putting on board the best Steam Engine with every Improvement with which he is acquainted or can devise he would be satisfied with it after the Boat had cleared all her expenses."[54]

Evans also invested his engines in operating concerns. Of the twenty engines he had built by 1812, he owned all or part of six: five used for mills and one in a steamboat. For those he sold outright, Evans gave terms of half down and the other half on interest. Since some of his engines were sent long distances and to areas where trained mechanics were not readily available, Evans also worked out a system of insurance. An engine of 72 horsepower, purchased with no insurance, sold in 1812 for six thousand dollars. For ten thousand dollars, however, Evans would guarantee the engine's operation for four years and grant an exclusive franchise for its use within a reasonable area. Under this arrangement, expenses for repairs during those years were to be deducted from the usual one thousand dollar yearly payments.[55] Evans, like the New York makers, had a large general engineering business to supplement his engine-building enterprise.

Another practice designed to reduce the capital needed to engage in engine building was that of requiring an advance from buyers. Latrobe wrote in 1813 that "in no case whatever that I have heard of, has an Engine been built without an advance of part of the cost. Even Mr. Fulton," he added, "who can better than any one else afford it, is paid by our Steam boat company at the rate of 2000£ month while going on with the work."[56] Such a practice, whether as universally practiced as Latrobe claimed, is both consistent with and indicative of the marginal nature of these early engine shops. This procedure had the secondary advantage of committing a buyer to carry out the transaction.

Skilled workmen, like sufficient funds, were scarce for early engine builders. Latrobe maintained that any workmen "who have general talents" could, with proper supervision and the help of the necessary smiths, build a steam engine, "though they never saw an engine in their lives." He added that "with such men I built the Waterworks in Phila[delphia]."[57] With such men, Fitch built his steamboat. Even with this possibility conceded, such operations were fraught with failure. When asked in 1811 whether he would become a manufacturer of steam engines, Latrobe replied emphatically that "I am the last Man in the U. States who would think of engaging in it for this reason, that having been the first to introduce their general use on this side of the Atlantic, no one is better acquainted than I am, with the excessive plague, & trouble, & disappointment of obtaining & keeping good workmen."[58]

Always enterprising, Evans seems to have been the most active in recruiting young workmen. In 1810 he advertised for "YOUNG MILLWRIGHTS" and held out to applicants the prospect of a bright future. They "may receive good wages," he promised, ". . . and be instructed in the art of making Steam Engines, building Steam Mills and attending them, and when capable, they will have opportunities of going out with Steam Engines; to Build Steam Mills, and to instruct others to keep them in repair, and attend them, for which they will be paid very high wages."[59] Unlike so many masters, Evans seems to have actually taught, rather than merely exploited, his young workmen. P. R. Hodge, the author of the excellent book *The Steam Engine, Its Origin and Gradual Improvement* (1840), was employed in the Mars Works as a young man. Two of the workmen that Evans sent to Pittsburgh later became important engine builders in their own right.

Most of these first engine shops had at least one trained English workman whose knowledge acted as a leaven among the rest of the men. In addition, workmen were available who had already gained some experience through previous attempts to build steamboats or to erect engines such as those for the Philadelphia Waterworks. Some workmen seem to have been in the employ, at one time or another, of several engine builders. One such workman named Baker, who had perhaps worked for Roosevelt at Soho, was sought simultaneously by Evans, Latrobe, and Fulton. It is significant that he was allowed to name his own terms of employment, which turned out to be three dollars per day plus traveling expenses. Latrobe was particularly pained in 1814 to find one of his former workmen from the Philadelphia Rolling Mill working for Evans in Pittsburgh.[60] The shortage of trained workmen was one problem that was never solved, and trained artisans of any description were at a premium in America.

Few details have survived concerning the actual shops in which Evans, Smallman, and the other enginemakers worked. The earliest description is that of Roosevelt's Soho Works on the Passaic River, in New Jersey. Thomas P. Cope, a representative of the Philadelphia watering committee, visited these works in July 1800 to check on the progress being made in constructing the large Philadelphia engines. According to his description, the pertinent parts of the works included a smith's shop 90 by 40 feet with six fires and two air furnaces; an adjoining room, 30 by 20 feet, had a fire for heavy work where iron parts were heated and wrought into appropriate size and shape, and where small castings were made. Another building, called the fitting shop, contained water-driven lathes and drilling machines, used principally to bore cannon. An adjacent shop contained another water-powered machine for boring even larger cylinders. Farther up the stream were the large blast furnaces where metal was cast directly from ore. Here also was a pattern shop with two foot-driven lathes, probably for turning wooden moulds.[61]

During his visit Cope watched the boring of one of the Philadelphia engines. The cylinder was 6 feet 6 inches long with an internal diameter, before boring, of 38¼ inches. The boring was to increase the internal diameter by three quarters of an inch. When Cope visited Soho, an increase of half an inch had been accomplished in 87 days! The drilling machine headpiece, slightly smaller than the bore of the cylinder, was fitted directly to the waterwheel shaft. Into this headpiece were fitted three steelings (cutters), each 3½ inches wide. The boring continued night and day, stopping every ten minutes to change the steelings, a process which itself took ten minutes. One workman "almost lives in the cylinder," Cope reported. This mechanic had a hammer and it was his duty to "keep things in order" and to attend to the steelings. A second man was in attendance to move the frame upon which the cylinder rested. When the work on this particular cylinder was finally done, it was discovered that the bore was, for all the pains taken, at least three-eighths of an inch greater in diameter at the center than at the ends.[62]

The Mars Works of Oliver Evans was nearly as complete as that at Soho. In 1808 it consisted of five basic shops used for making engines: an iron foundry; a pattern shop; a blacksmith's shop; a steam mill for turning and boring; and a "Steam Engineer's Shop," where the finished product was assembled.[63] Fulton's shops at Jersey City were similarly extensive. His engine works was described as "a capacious building, two stories high, in which are the boring, turning, and drilling mills; also, the fitting, filing, and model shops. The second building is a superb smith's shop, containing nine fires, in which shafts of one-ton weight are forged. The third is a *boiler* shop, with its *oven, moulds, punchers,* and *cutters,* complete."[64] The separate boiler shop was a feature not always found in other engine shops. By having his castings done by McQueen and Allaire, he was able to dispense with the usual air furnace and mould shop.

The most minutely described engine works is the one built by Latrobe at Pittsburgh in 1814. The long, two-story building was made of rough boards. A hearth and two lathes were in the filer's shop; the smith's shop contained four "common" and one large fire. The boiler shop had "two hearths, a drill frame, punching frame & sheers, & a furnace for bending boiler plates." Each shop also had a complete set of tools. "I have in the center of my works," Latrobe explained, "an upright shaft, and Horse walk, which turns my Lathe's & drills.[65] This equipment was calculated to be powered by four horses, but Latrobe claimed that he never used more than one.

Next to finding a furnace to cast the cylinders, proper boring was perhaps the biggest problem of the enginemakers. Latrobe affirmed that "the story of Cylinders not being capable of being as well & truly bored here as in England has no

THE LATROBE ENGINE WORKS, PITTSBURGH—1814[68]

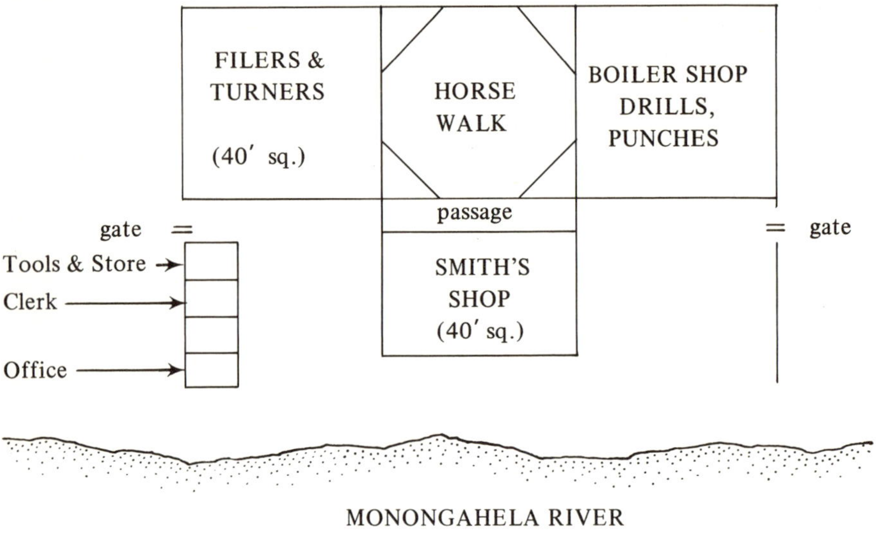

foundation." The Navy Yard engine, made by Smallman, had "a cylinder as perfectly bored as ever was bored in England."[66] William Thornton was less sanguine. "I was not acquainted with Mr. Smallman's Boring machine till lately described," he wrote, "and fear that though it may bore very well it will not give that finish that a good Cylinder ought to have, for it should be highly polished."[67]

David Wilkinson, the Rhode Island machine maker and former partner of Elijah Ormsbee, visited the steamboat shops of Fulton and John Stevens and inspected the latter's boring mill. "I thought he was ahead of Fulton, as an inventor" Wilkinson later declared.[69] Between 1805 and 1808 Wilkinson took part in the manufacture of cannon at Pawtucket. The cannons were cast solid, and bored out by a unique water-powered machine which held the drill stationary and rotated the casting.[70] The boring of cylinders generally was patterned closely after the boring of cannon, an art not unknown even in colonial America. Continuing military action against Indians in the West, for example, stimulated the growth of a weapons industry in that area which in turn had a positive effect on the growth of engine making.

When John Stevens attempted to justify his failure to build a steamboat in North Carolina, he wrote that "a vessel might have been built in any part of that state, but it would have been almost impracticable to construct the engine any other place than New York or Philadelphia."[71] It was in these cities that steam-engine builders could find whatever advantages America held for them before 1812. Here were the

TOOLS LISTED FOR LATROBE ENGINE WORKS–1814[72]

Smith's Shop

4 large cast anvils
4 wrought do.
5 complete sets of smith's tools adapted to all the purposes of erecting engines
2 large cranes and hangings
9 cast dies [?] and moulds for jaws and joints
3 pair of common and 3 of large smith's bellows
1 large and 1 smaller vice
1 beak iron
1 complete set of nail and spike tools

Filer's Shop

2 large capital lathes
1 small do. and man wheel
6 vices
2 press frames
1 pair bellows
1 complete assortment of files, cutters, screw-plates, drill, turning tools etc.

Boiler Shop

2 pairs smith's bellows
2 anvils
1 punch block, dies, etc.
1 pair shears
1 drill frame, turned by horse, drills, and reemers
3 pair cast iron boiler plate moulds
1 complete set of drift wedge and screw bolts, tongs, etc.

Millhouse

A perpendicular shaft, wheels, and drums to turn the lathes and drills

Pattern Shop (upstairs)

1 lathe for the pattern maker
6 carpenter's benches and screws
2 cast iron stoves
A number of patterns for the engines

large populations which could supply artisans, finance, and the patronage of large public works. Here, too, were centers of manufactures and transportation, which represented large potential markets for steam engines. In Philadelphia, Smallman, Evans, Large, and other minor and transitory enginemakers set up the first shops of their kind in the country. Fulton, McQueen, and Allaire quickly followed suit in New York. Before the outbreak of war in 1812, these two cities had firmly established themselves as the major suppliers of steam engines to the new nation.

Notes

[1] Manuscript notes, not paged, in University of Michigan, Transportation Library copy of Oliver Evans, *The Abortion of the Young Steam Engineer's Guide* (Philadelphia, 1805).

[2] Thomas Green Fessenden, *The Register of Arts, or A Compendious View of Some of the Most Useful Modern Discoveries and Inventions* (Philadelphia, 1808), page 333.

[3] Nicholas Roosevelt to John H. B. Latrobe, 29 March 1828, Fulton papers, NYHS.

[4] Deposition of James Smallman, 24 November 1814, quoted in Thompson Westcott, "Steam Navigation, in 1798. Rooseveldt: Livingston: Fulton," *Historical Magazine,* volume 1, 3d series (January 1872), page 24.

[5] B. H. Latrobe to Eric Bollman, 6 February 1804, B. H. Latrobe MS. Coll., MHS.

[6] B. H. Latrobe to Charles Gobert, 16 March 1804, B. H. Latrobe MS. Coll., MHS.

[7] B. H. Latrobe to James Smallman, 5 October 1810, B. H. Latrobe MS. Coll., MHS.

[8] B. H. Latrobe to Secretary of the Navy, 1 July 1810, B. H. Latrobe MS. Coll., MHS.

[9] William Thornton to James Van Ingen, 14 December 1810, Thornton papers, volume 3, LC.

[10] B. H. Latrobe to Samuel Hazelhurst, 3 February 1815, B. H. Latrobe MS. Coll., MHS.

[11] Woodbury Report, *Letter from the Secretary of the Treasury, transmitting . . . information in relation to steam-engines,* &c, pages 255-256.

[12] B. H. Latrobe to James Smallman, 25 January 1811, B. H. Latrobe MS. coll., MHS; William Thornton to James Van Ingen, 14 December 1810, Thornton papers, volume 3, LC.

[13] B. H. Latrobe to William Chamberlain, 8 February 1811, B. H. Latrobe MS. Coll., MHS.

[14] Returns of Marshalls re: Enemy Aliens, 1812, record group 59, records of the Department of State, National Archives. Cf. a passport made out to a Daniel Large in the Library Company manuscript collections, Philadelphia, Pennsylvania; this engine was still operating in 1838. Woodbury Report, page 159. An engine built in 1804, and attributed to Large, was also in operation. Ibid., page 163. Cf. Miriam Hussey, *From Merchants to "Colour Men": Five Generations of Samuel Wetherill's White Lead Business* (Philadelphia, 1956).

[15] B. H. Latrobe to William Davy, 28 July 1811, B. H. Latrobe MS. Coll., MHS.

[16] B. H. Latrobe to James Smallman, 28 July 1811, B. H. Latrobe MS. Coll., MHS; B. H. Latrobe to Daniel Large, 26 October 1811, B. H. Latrobe MS. Coll., MHS.

[17] Daniel Large and Francis Grice to William Thornton, 19 December 1812, Thornton papers, volume 4, LC.

[18] *American Daily Advertiser,* 27 November 1816, quoted in Preble, *Chronological History,* page 88.

[19] Fritz Redlich, "William Jones and His Unsuccessful Steamboat Venture of 1819," *Bulletin of the Business History Society,* volume 21 (November 1947), pages 125-136.

[20] Westcott, "Steam Navigation," *Historical Magazine,* volume 1 (1872), page 23.

[21] B. H. Latrobe to James Smallman, 2 July 1813, B. H. Latrobe MS. Coll., MHS.

[22]Evans, *Abortion,* pages 95-96.

[23]See description and illustration in undated Addenda to *Niles' Weekly Register,* volume 3 (September 1812-March 1813), pages 1-2.

[24][Oliver Evans], *Oliver Evans and His Counsel, Who Are Engaged in Defense of His Patent Rights* . . . (no place, [1816], page 33 n.

[25]Evans, *Abortion,* page 97.

[26][Oliver Evans] Patrick N. I. Elisha, *Patent Right Oppression Exposed; or, Knavery Detected* (Philadelphia, 1813), pages 170-171. Donaldson, a New Orleans merchant, reportedly wanted "a Steam Engine & mill to saw timber with six saws at once." John Francis McDermott, editor, "The Western Journals of Dr. George Hunter, 1796-1805," *Transactions of the American Philosophical Society,* volume 53, part 4 (1963), page 118. Donaldson ordered the engine through Hunter.

[27]*Aurora,* 26 February 1803.

[28]Ibid., 27 September 1803.

[29]Ibid., 5 March 1804; Bathe, *Evans,* pagc 67fn.

[30]*Aurora,* 15 July 1805.

[31]Ibid., 30 April 1807.

[32]Ibid., 2 June 1807.

[33]Ibid., 19 November 1808.

[34]Memo dated 21 May 1809, among manuscript notes in University of Michigan copy of Evans, *Abortion.*

[35]Oliver Evans, "Account of the Steam Engines of Oliver Evans," *Archives of Useful Knowledge,* volume 2 (April 1812), pages 363-366.

[36]*Pittsburgh Gazette,* 18 December 1812.

[37]*Niles' Weekly Register,* volume 46 (5 July 1834), page 315.

[38][John Adams Dix], *Sketch of the Resources of the City of New-York* (New-York, 1827), page 44.

[39]B. H. Latrobe to Nicholas Roosevelt, 16 March 1804, B. H. Latrobe MS. Coll., MHS.

[40]Latrobe, "First Report," *Transactions of the American Philosophical Society,* volume 6, part 1 (1804), page 91.

[41]Robert Fulton to Boulton and Watt, 4 November 1794, in Discussion on Charles H. Haswell, "Reminiscences of Early Marine Steam Engine Construction and Steam Navigation in the United States of America from 1807 to 1850," *Transactions of the Institution of Naval Architects,* volume 40 (London, 1898), page 111; Robert Fulton to Boulton and Watt, 6 August 1803, ibid., pages 111-112.

[42]See table in Cadwallader D. Colden, *The Life of Robert Fulton* (New York, 1817), facing page 259. For the new shops see *American Weekly Messenger,* volume 1 (16 October 1813), page 64.

[43]B. H. Latrobe to Henry Latrobe, 18 September 1815, B. H. Latrobe MS. Coll., MHS; B. H. Latrobe to Frederick Graf, 27 August 1815, B. H. Latrobe MS. Coll., MHS.

[44]B. H. Latrobe to Henry Latrobe, 8 November 1815, B. H. Latrobe MS. Coll., MHS.

[45]Talbot Hamlin, *Benjamin Henry Latrobe,* page 472n; Morrison, *Steam Navigation,* page 44; schedules of the 1820 census of manufactures, New York, no. 1282, record group 29, National Archives.

[46]Morrison, *Steam Navigation,* page 42; Brown, *Allaire's Lost Empire,* pages 19-20; schedules of the 1820 census of manufactures, New York, no. 1292, record group 29, National Archives.

[47]*Appletons' Cyclopaedia of American Biography,* volume 5, page 268; ibid., volume 2, pages 122-123; ibid., volume 3, page 113; Nelson P.

Lewis, "Charles Haynes Haswell," *Transactions of the American Society of Civil Engineers,* volume 61 (1908), pages 553-555; Haswell, *Reminiscences,* page 269.

[48] Robert Fulton to Oliver Evans, 3 January 1812, Dreer collection, inventors, Historical Society of Pennsylvania, Philadelphia, Pennsylvania.

[49] Deposition of James Smallman, 24 November 1814, in Westcott, "Steam Navigation," *Historical Magazine,* volume 1 (1872), page 24.

[50] John Fitch to Dr. Madison, 10 February 1788 (?), Fitch papers, volume 2, LC.

[51] B. H. Latrobe to William Hartshorne, 6 July 1813, B. H. Latrobe MS. Coll., MHS.

[52] B. H. Latrobe to William Chamberlain, 8 February 1811, B. H. Latrobe MS. Coll., MHS.

[53] William Thornton to James Van Ingen, 2 December 1810, Thornton papers, volume 3, LC.

[54] William Thornton to John Galvin, 14 December 1810, Thornton papers, volume 3, LC.

[55] *Niles' Weekly Register,* volume 3 (17 October 1812), page 111.

[56] B. II. Latrobe to Alcxandcr McKim, 20 March 1813, B. H. Latrobe MS. Coll., MHS.

[57] B. H. Latrobe to Secretary of the Navy, 26 September 1811, B. H. Latrobe MS. Coll., MHS. Of the five men sent out by Smallman to put up the Navy Yard engine in 1810, two were millwrights and two were smiths. B. H. Latrobe to Secretary of the Navy, 1 July 1810, B. H. Latrobe MS. Coll., MHS.

[58] B. H. Latrobe to Robert Smith, 5 August 1811, B. H. Latrobe MS. Coll., MHS.

[59] *Aurora,* 15 August 1810.

[60] B. H. Latrobe to Robert Fulton, 28 October 1812, B. H. Latrobe MS. Coll., MHS; B. H. Latrobe to Henry Latrobe, 13 November 1814, B. H. Latrobe MS. Coll., MHS.

[61] Description of Thomas P. Cope, 4 July 1800, quoted in "History of the Steam Engine," *Journal of the Franklin Institute,* volume 102 (1876), page 259.

[62] Ibid., pages 259-260; Graff, "Earliest Steam Engines," *Journal of the Franklin Institute,* volume 55 (1853), page 271.

[63] *Aurora,* 19 November 1808.

[64] *American Weekly Messenger,* volume 1 (16 October 1813), page 64.

[65] B. H. Latrobe to William Stackhouse, 24 May 1814, B. H. Latrobe MS. Coll., MHS.

[66] B. H. Latrobe to Andrew Hazelhurst, 19 October 1810, B. H. Latrobe MS. Coll., MHS.

[67] William Thornton to James Van Ingen, 14 December 1810, Thornton papers, volume 3, LC.

[68] After a drawing by B. H. Latrobe, in a letter to Robert Fulton, 21 May 1814, B. H. Latrobe MS. Coll., MHS. A pattern and ship-joiner's shop were located on the second floor of the long building.

[69] Wilkinson, "Reminiscences," *Transactions of the Rhode Island Society for the Encouragement of Domestic Industry . . . 1861,* pages 104-105.

[70] Edward S. Wilkinson to Elisha Dyer, 16 December 1861, in *Transactions of the Rhode Island Society for the Encouragement of Domestic Industry . . . 1861,* page 89. John Wilkinson, of Bradley, patented this type of boring machine in England in 1774. The cylinders for most of Boulton and Watt's early engines were bored with this apparatus.

[71] Quoted in Turnbull, *Stevens,* page 344.

[72] B. H. Latrobe to Robert Fulton, 21 May 1814, B. H. Latrobe MS. Coll., MHS.

Chapter 5

The Spread of Steam Power

PITTSBURGH, in this branch of business, has acquired great celebrity. . . . It may be said, without exception, that there is no place in the world, that can surpass Pittsburgh, as to the means and materials, which it affords for the manufacture of these powerful machines. The talent and knowledge, too, are possessed by our mechanics in all necessary perfection.

SAMUEL JONES (1826)[1]

On the eve of the War of 1812, the great trans-Appalachian West was just beginning that unprecedented expansion which—today as then—beggars the imagination. The population west of the mountains and north of the Ohio River stood at 51,000 in 1800. In 1810 it had risen to 292,000, and by 1820 it had grown to 859,000—more than seventeen times what it had been twenty years before. Cities were built, internal improvements were undertaken, manufactures grew, and land values soared in the froth of this migration. Almost from the beginning, the steam engine played a major role in the great adventure. If steam power was a convenience in the East, it was indispensable in the West. No other power was available to stem the currents of the broad rivers or to repair the absence of millstreams.

Yet what the West demanded in the way of power, it was able to provide. In no other part of the United States were so many advantages available for the making of engines and the use of steam: iron ore in easy abundance; charcoal for smelting and stone coal for boilers, almost for the picking up and burning; fresh water for any number of boilers; and a steady influx of artisans and craftsmen. Benefits were: for the user, a comparatively cheap source of engines; for the maker, a growing and clamorous market. The engineer and manufacturer quickly seized upon the steam engine as an instrument of great utility and certain profit.

Pittsburgh was the first center of steam engine building and utilization in the West. All the western advantages seemed epitomized in this raw town at the confluence of the Ohio, where the Monongahela and Allegheny Rivers merged. Coal and iron were close and abundant, a growing population provided skilled workmen, and new manufacturing establishments provided a market. Just as engine building along the east coast had shown a remarkable continuity with English technique and engineers, so the early western shops first appeared as mere extensions of eastern activity.

A general demand for cast-iron products preceded the specific market for steam engines. The first iron foundry in the town was completed in 1806 by Joseph McClurg. Called the Pittsburgh Foundry, it was expanded during the War of 1812 to produce cannon, and a large order for it was secured by McClurg.[2] Anthony Beelen opened his Eagle Foundry in Pittsburgh in 1810.[3] The establishment of these facilities was important, because it meant that prospective engine builders could begin working without having to invest immediately in their own foundries. Discussing engine building in 1814, Benjamin Henry Latrobe pointed out that while Wheeling, Virginia, had certain advantages of location, Pittsburgh had "the advantage of foundries & extensive establishments on every branch of the Iron business." No place westward, he added, could offer such facilities.[4]

The first steam engine built in Pittsburgh was one for the Fulton steamboat company in 1811. Nicholas Roosevelt and Charles Stoudinger were sent west to supervise the work on this vessel.[5] Not long after, Daniel French, who had been so interested in the Elijah Ormsbee steamboat, began building steamboats near Pittsburgh. French received a patent for his improved steam engine in 1809 and before 1811 had built two steamboats in New York.[6] In 1811 he transferred to Philadelphia to exploit his newly invented brickmaking machine. While there he built at least one steam engine—for William Cooper's ferry on the Delaware River.[7] From 1812 to 1814, French built several small steamboats at Brownsville, on the Monongahela near Pittsburgh.[8]

Oliver Evans first extended his business interests to Pittsburgh in 1809 when in partnership with son George and others, he established a steam-driven flour mill. His analysis of the industrial potential of the area led him to establish an engine manufactory there in 1812. In the summer of that year, he announced that he had decided to set up his son George in the engine business to protect his interests because "certain steam engineers of the Borough of Pittsburgh pretend to claim a right to make and sell my patented steam engines."[9] By December, George Evans announced that he was "ready to contract for the supplying of STEAM ENGINES on Oliver Evans's principles."[10]

Evans' Pittsburgh Steam Engine Company, like his Mars Works in Philadelphia, carried on a business in general iron and millwork trade. Besides steam engines, they made "all kinds of Machinery; together with wrought Iron Work, and Castings, &c. &c. Such as all kinds of wrought and cast Iron. Work appertaining to Merchant Flour Mills; Rolling and Slitting Mills; Forges; Tilt Furnaces; Sugar, Paper, and Saw Mills, &c. and all work of a similar kind; together with every other description of castings, that may be wanted." In addition, the company offered to "also make patterns, and do boring and turning in Iron, &c."[11] Three years later this manufactory, "embracing a foundry and smithery," was employing a hundred men and reportedly had made many of the anchors for Oliver Hazard Perry's fleet on Lake Erie.[12]

The westward growing rank of steam engineers was further swelled in 1813 by the arrival of Benjamin Henry Latrobe. Latrobe had built a small engine shop in Washington in 1811 to hasten the completion of the engine for the New Orleans Waterworks. The British blockade in the War of 1812, however, played havoc with communication by sea with New Orleans, as well as with all coastal trading between Virginia and New York. In August 1811 he had considered the possibility of establishing an engine works in Pittsburgh, but decided that such an enterprise would not then be profitable. He had no doubt that steam engines would take the place of water mills in the West. "But the demand for Steam engines," he warned, "must precede the attempt to manufacture them, and that demand can only grow out of the increasing population." Six large engines, he guessed, would "grind up the whole country."[13]

Two subsequent developments led Lathrobe to change his mind: the growing difficulty of communicating with New Orleans by sea and the possibility of building a large number of small engines to be used on sugar plantations in Louisiana. A planter from that state appeared in Washington in 1812, asking about the terms upon which a sugar mill engine could be procured. Latrobe estimated that $2,500 would be a fair price for a small 12-inch cylinder engine.[14] At this low price he did not plan to make a profit on the first engine. Five or six hundred dollars of this price would be absorbed in "making patterns for castings, & the necessary tools & arrangements for an Engine of that size." It was his hope that the contemplated engine would lead to further orders in the South. If he could receive such additional orders, the five or six hundred dollars for patterns and tools could be saved in the building of each new engine and would constitute a satisfactory profit.[15]

If such orders were forthcoming, the establishment of a manufactory in Pittsburgh became more reasonable. His previous scepticism had been based on the fear that the immediate area would not provide a sufficient market for engines. The development of a market in the South meant that engines could be made in Pittsburgh, where materials and workmen were obtainable, and then shipped down the river at very little expense. Latrobe's decision to move west was strengthened by news that Robert Fulton intended to remove his main steamboat interests from New York to western waters. He feared that engine building facilities in Pittsburgh were not sufficient to support two large engine works, but hoped that through cooperation both he and Fulton might prosper.[16]

Latrobe, with the hope of soon being "over whelmed with applications" for engines from Louisiana, suggested to Fulton that the two of them join forces. Both men were to operate in the same manufactory and use the same workmen. Latrobe would act "as a professional man" in supervising the building of Fulton's steamboat engines without having a financial interest in the Fulton company. Fulton, on the other hand, would be a full partner in the Latrobe engine-building firm. The anticipated benefits were that Fulton would gain the use of an expert engineer, Latrobe would have the financial backing of Fulton, the shops could continue to make sugar-mill engines after the contemplated steamboats were completed, and both men would make a handsome profit from the southern trade.[17] Under the impression that these arrangements were acceptable to Fulton, Latrobe sold his small Washington shop and proceeded to Pittsburgh. He arrived in October 1813 and began immediately to employ workmen that he placed temporarily under Charles Stoudinger at the Fulton works.[18]

Upon his arrival in Pittsburgh, Latrobe made arrangements to operate out of the steam-engine shops of Thomas Copeland until other facilities were available.[19] Copeland had announced his readiness to build engines "upon the principles of the celebrated WATT & BOLTON" in April 1813. "Being prepared with the necessary machinery and tools of every kind for the execution of all parts of an engine," he was ready to "engage with any person to make and set up an engine of any power that may be required." He further maintained that since the elastic power of steam was a well-known fact of nature, no pretended monopoly would prevent him from using "steam of any elastic force."[20]

This direct assault on the Evans interests, as they were conceived, did not go unchallenged. Adopting the invective style of the day, George Evans expressed confidence that "most men of talents to amass the means of putting into operation a Steam Engine and its appurtenances to business, which will cost from 2 to 40 thousand dollars, are also competent to expend their wealth prudently." Asserting that "not one in a thousand has any knowledge" of the subject of steam, he proposed to let the controversy die.[21] Copeland insisted upon having the last word and assured the public that he had a "steady determination to adhere to his proposals." In order to forestall fears of any lawsuits which Evans might instigate, he further offered to "warrant against all pretended claims for damages which may be made by any patentee."[22]

The fact that Copeland was willing to rent shops to Latrobe in the winter of 1813 indicates that at first he was not too successful in securing orders for his engines. Early in the spring of 1814, however, he received orders for three engines and asked Latrobe to vacate the premises.[23] One of these engines, with a 24-inch cylinder, was being built for a Mr. Brady of St. Louis, priced at $8,200 plus transportation. The other two engines, each with 20-inch cylinders, were ordered by mills in Steubenville, Ohio, at a price of $6,000 each. Copeland confessed that these prices were at cost, and presumably he accepted the orders merely to get his own enterprise started.[24] By 1815 Copeland was said to be "extensively" engaged in engine building.[25]

Forced to vacate Copeland's shops, Latrobe built his own. Here he simultaneously attempted to build a steamboat for Fulton and to construct several engines for his own use, including the one for the New Orleans Waterworks. With the exception of an engine built on the plans of Francis B. Ogden for a woolen mill in Steubenville, none of these tasks was completed. Financial misunderstanding with Fulton, who claimed not to have authorized the building of new shops, was the major cause of failure. When Latrobe was offered his old government position as surveyor of public buildings in Washington, he left Pittsburgh for good.[26]

Another engine shop was added to those already operating in Pittsburgh when James Arthurs began to build engines of the Boulton and Watt type in 1815. Arthurs had been operating a wool-carding mill as early as 1812, and by 1814 or 1815 he was using a steam engine to operate some of his expanded facilities.[27] In

1815 Evans brought action against him for making engines that used steam of more than atmospheric pressure, although he did not consider Arthurs' business to be a very great threat or the builder to be a very "powerful adversary."[28] The Arthurs textile mill was still operating in 1826, as was the engine works until 1836, under the proprietorship of John Arthurs.[29] In 1832 the latter establishment was producing "steam engines and castings for machinery of all descriptions," employing forty workmen and making forty thousand dollars worth of ironwork a year.[30]

When Oliver Evans died in 1819, the partners, George Evans, Mahlon Rogers, and Luther Stephens (as well as Mark Stackhouse, the head workman), who operated the Pittsburgh Steam Engine Company, dissolved their concern and went separate ways. With the exception of Luther Stephens–who had already returned to Lexington, Kentucky–the main figures of the firm remained in the engine-building business. Besides continuing his steam gristmill, George Evans also seems to have retained some interest in the steam engine factory.[31]

Mark Stackhouse had been employed by Evans for many years. In 1802 he was sent from the Mars Works to help set up the first Evans engine installed in the West. He was with the Pittsburgh Steam Engine Company from the beginning, probably as manager of the engine building. He set up his own factory in 1819 and by 1826 it was one of the largest in the West. His Columbia Steam Engine Company, perhaps typical of its size, was turning out seven engines a year, none under 60 horsepower, at an average price of thirty thousand dollars. Connected with this factory, Stackhouse had a turning, boring, and grinding establishment powered by steam and doing a general business in custom millwork. With a partner he also operated another steam works turning out engines for boats. This factory too had a foundry attached.[32]

Mahlon Rogers appears to have been a workman at Latrobe's Philadelphia rolling mill about 1804 and was a partner in the Pittsburgh Steam Engine Company from its inception. He may have had charge of installing engines in manufactories such as Cowan's Rolling Mill in Pittsburgh. Evans' largest engine up to that time, one of 70 horsepower, ran this mill. When Latrobe visited the works in 1814 he complained that it was "an exact copy of mine on the Schuylkill" and that Cowan's head roller, as well as Rogers', had worked for him previously in Philadelphia.[33]

Few people in Pittsburgh were greatly concerned with the niceties of mill design or the advantages that the enginemakers took, each of the other. The Cowan engine itself, however, was a cause for wonder and rejoicing. In the 1880s an old man remembered his boyish excitement at

> the erection of this mill, and the important day on which the engine was to be put into operation. A large number of citizens had assembled to witness the great event. The engine was an upright cylinder, with a large log for a walking beam. At noon the fire was put under the battery of boilers, and some time elapsed before enough steam was generated to warrant an attempt at starting. When the time came and the engineer let on the steam, there was much whistling and blowing, but no answering motion came from the great stubborn flywheel. A call for help was made, and the

bystanders armed themselves with levers and handspikes and went at the wheel to force it over the centre. It was some time before that purpose was accomplished, but after repeating it several times it went off as a thing of life, and the engine of Pittsburgh's first rolling-mill was sturdily at work. The engineer permitted some of us boys to climb on the walking beam, and we went up and down, highly delighted with our novel ride.[34]

After Evans' death in 1819 Rogers operated his own engine factory. In 1832 he was producing ten thousand dollars worth of engines a year, with an average of 18 workmen. While his business was small at this time, it should be noted that he built engines exclusively, rather than carrying on a general trade as did most of his competitors.[35]

Sometime after Latrobe visited the Pittsburgh Rolling Mill in 1814, that establishment too became a steam engine factory. During the winter of 1814-15, Latrobe acted as agent for William Stackpole of Boston when the latter bought the works from Cowan.[36] The Pittsburgh Iron, Nail Factory, as it was then styled, began making engines in 1818 and immediately ran into opposition from Oliver Evans. J. Whiting, agent for the firm, maintained that they were confining themselves "to the principles, machinery and general construction invented and used in England by Watt and Bolton, by Freemantle, by Trivithick, Vivian and various other *English* engineers."[37] George Evans claimed that "Whiting does make such Steam Engines as come within Oliver Evans's patent," adding that "we have not known him ever to make any other kind."[38] The fact was that Whiting used an engine of the Watt type and high-pressure steam at the same time, a not uncommon arrangement and one to which Evans always objected. The controversy ended in 1819 with the death of Evans and the bankruptcy of Stackpole and Whiting.

In 1832 the local engine builders were joined by John Snowden, Jr., who erected a shop in nearby Fayette County. This shop did engine and general ironwork and employed twenty men.[39] Of more importance was the Penn Foundry and Steam Engine Manufactory of McClurg, Pratt and Wade, located in Pittsburgh. The foundry had been the earliest in the West, and the engine shop was added in 1830. One of the most extensive establishments in the West, it produced in one year 14 engines, 140 tons of cannon, and 800 tons of other castings for an aggregate value of $95,000. A part of this production was sold in the vicinity of Pittsburgh, the remainder was "sent to Philadelphia, east, and from the lakes to the mouth of the Mississippi, west," transactions usually being on six months' credit. Some competition from foreign-built engines was reported from New Orleans, but there was apparently none locally.[40]

As the population of Louisville grew, like Pittsburgh, it too began to produce steam engines. A correspondent in Baton Rouge wrote Evans in 1807 that he had "learned that 2 or 3 Steam Engines had been ordered from Kentucky, of the price of 500 dollars."[41] This low price casts doubt on the accuracy of the information. The first iron foundry in Louisville, where engine manufacture might reasonably be expected to start in the state, was not built until 1812, by Paul Skidmore. The latter sold out in 1814 or 1815 to Joshua Headington, who in turn sold the works to David Prentice in 1816.[42]

Prentice was a Scottish millwright who arrived in this country sometime before 1811. In March 1813 he was operating the Eagle Works (formerly Foxall's), a foundry on the Schuylkill near Philadelphia.[43] By this date he had already built two steamboats and was working on another.[44] When Jacob Perkins, the inventor from Newburyport, Massachusetts, sought an evaluation of Evans' engine from Latrobe, the latter referred him to Prentice as the only man who had used and made a close comparative study of both types of engine.[45] Prentice had purchased an Evans engine and operated it with the same boiler that he had used for an engine of the Boulton and Watt type.[46]

Latrobe at this time asked Prentice to make a cylinder to be sent to Washington, explaining that the British blockade of the Chesapeake made it impossible for him to patronize his regular iron founder.[47] Prentice, however, was in similar trouble and reported to his father that "business is getting worse with me on account of the unhappy war." Stone (bituminous) coal from Richmond, Virginia, upon which Philadelphia usually relied, was cut off by the blockade and had risen in price from 40 to 75 cents a bushel. Even at these inflated prices, the supply was nearly depleted in Philadelphia in March 1813. If the blockade lasted, Prentice feared it would be necessary for him to "close up the casting altogether."[48]

Another wartime disruption was caused by the search by the United States Marshall of Pennsylvania for Irishmen suspected of subversion and treasonous inclinations. Prentice employed a number of alien Irish and the threat of their deportation, or removal forty miles from tidewater, further demoralized the foundry.[49] In the face of such difficulties, Prentice accepted the offer to transfer to Kentucky to erect a steam sawmill and gristmill for an old friend. While so engaged, he also found time to build a steamboat on the Ohio.[50]

Prentice purchased the foundry in Louisville in 1816 and six months later was joined in partnership by Thomas Bakewell, an English-born friend of the Darwin family, brother-in-law to James Audubon, and son of the man who had first helped Prentice set up in business in Philadelphia.[51] Bakewell joined the firm without funds, having sold his steam mill on credit to Audubon. The two partners, nevertheless, undertook the manufacture of the first steam engines in Louisville. The dangers of beginning to build engines without sufficient capital were so great that Bakewell later wrote: "most of my dreams even now (1873) revert with painful vividness to that period."[52]

Toward the end of the partnership, in 1821, their engine works was doing a large amount of work. In one year they built eight engines for steamboats and two for stationary use, with an aggregate value of one hundred thousand dollars. In addition, their foundry turned out another five thousand dollars worth of general castings for the surrounding countryside. In 1819 when these figures were quoted, the works had another seventy thousand dollars worth of engines and steamboats under contract.[53] After the demise of the Prentice and Bakewell partnership in 1821, the latter moved to Cincinnati and continued to build and to improve steamboats and other machines.[54]

So rapid was the growth of engine building in Louisville, that between 1819 and 1838 at least ten different engine firms had together built 246 engines: 65 for steamboats and 181 for mills of various kinds.[55] This record is remarkable when one considers that in 1819 Louisville had a population of only 4,500.

Although the manufacture of steam engines in Cincinnati was begun later than in Louisville, the business grew almost as rapidly. The town had no iron foundry in 1815, but it was reported to be well supplied with blacksmiths.[56] The iron foundry of William Green was the first to operate in Cincinnati, and by 1818 it had eighty men at work on the construction of seven steamboats.[57] The following year a nail factory and fifty more hands were added to the establishment. By 1823 Thomas Bakewell, William Tift, and John A. Morton had joined the ranks of Cincinnati engine builders. In 1826 the city had five steam engine factories employing 126 men and producing $134,000 worth of engines in a single year.[58]

The nature of the Cincinnati engine shops in 1826 showed a wide diversity of production that was probably typical of most such enterprises in the West. For example, the Phoenix Foundry of C. Tatum and Sons built "various kinds of Machinery," such as mill castings. Goodloe and Harkness' "Cooper Foundry, Cotton Spinning, and Steam Engine Factory" operated 336 spindles besides making engines. Tift's Steam Engine and Finishing Establishment was housed in a new frame building, the second floor of which was intended "for the reception of an extensive cotton spinning establishment, which Mr. Tift, with his characteristic enterprize, is now forming."[59] The diversity of enterprise by these firms would seem to indicate that the number of engine builders had multiplied so rapidly that the business could not absorb the capital and talent available. It is safe to say that any engine factory could do general foundry work, and a great many foundries could also build engines.

In 1810 there were no steam engine shops and very few engines west of the Appalachians. A decade later, Pittsburgh, Louisville, and Cincinnati each had a plurality of such establishments, and the number of engines they produced was undoubtedly quite large. The steadily rising demand for steamboats to navigate the western waters accounts for part of this activity. Before the Civil War, no other part of the world built or used so many steamboats as did the western section of America. Another major factor was the growth of manufactures in the West, the general change of production from hand and home to machine and mill, and the mechanization of extractive industries in the South. Since waterpower sites were scarce and unreliable in most of the South and West, steam power made possible and profited from this increase in manufactures.

Underlying this growth was the rapid expansion of the population of the West during and after the War of 1812. The center of the nation's population, situated somewhere on the Eastern Shore of Maryland in 1790, had by 1840 moved to western Virginia (now West Virginia), just south of Pittsburgh. This influx of people demanded better transportation and more manufactured goods, and from it were drawn the skilled workmen to supply these needs. By 1820 the engine-making centers of the West were rivaling Philadelphia and New York in numbers and quality of engines built.

Notes

[1] S[amuel] Jones, *Pittsburgh in the Year 1826* . . . (Pittsburgh, 1826), page 59.

[2] *Documents Relative to the Manufactures in the United States* . . . , House Executive Documents Numbers 222, 223, 22d Congress, 1st session (Washington, D.C., 1833), volume 2, page 252, hereafter cited as McLane Report; Swank, *Iron in All Ages,* pages 226-227; Stephen L. Goodale, *Chronology of Iron and Steel* (Pittsburgh, 1920), page 83; B. H. Latrobe to John D. DeLacy, 25 December 1814, B. H. Latrobe MS. Coll., MHS.

[3] *Pittsburgh Gazette,* 14 December 1810.

[4] B. H. Latrobe to Michael Graham, 1 February 1814, B. H. Latrobe MS. Coll., MHS.

[5] B. H. Latrobe to Mrs. Roosevelt, 31 December 1810, B. H. Latrobe MS. Coll., MHS; *Pittsburgh Gazette,* 18 October 1811.

[6] Morrison, *Steam Navigation,* page 516; Daniel French to William Thornton, 12 February 1814, Thornton papers, volume 4, LC.

[7] Daniel French to William Thornton, 12 February 1814, Thornton papers, volume 4, LC.; B. H. Latrobe to William Cooper, 14 October 1812, B. H. Latrobe MS. Coll., MHS; Turnbull, *Stevens,* page 295. For French's brick machine see *Pittsburgh Gazette,* 14 February 1812.

[8] Daniel French to William Thornton, 12 February 1814, Thornton papers, volume 4, LC.; B. H. Latrobe to Robert Fulton, 1 March 1814, B. H. Latrobe MS. Coll., MHS; same to same, 1 May 1814, Latrobe MS. Coll., MHS.

[9] *Pittsburgh Gazette,* 24 July 1812.

[10] Ibid., 18 December 1812.

[11] Ibid., 9 April 1813.

[12] *Niles' Weekly Register,* volume 8 (29 April 1815), page 141.

[13] B. H. Latrobe to Robert Smith, 5 August 1811, B. H. Latrobe MS. Coll., MHS.

[14] B. H. Latrobe to Henry Latrobe, 22 March 1812, B. H. Latrobe MS. Coll., MHS.

[15] B. H. Latrobe to Mr. Coiron, 1 April 1812, B. H. Latrobe MS. Coll., MHS; B.H. Latrobe to F.C. Graf, 5 May 1812, B. H. Latrobe MS. Coll., MHS.

[16] B. H. Latrobe to Robert Fulton, 5 July 1812, B. H. Latrobe MS. Coll., MHS.

[17] B. H. Latrobe to Robert Fulton, 31 July 1812, B. H. Latrobe MS. Coll., MHS.

[18] B. H. Latrobe to Joseph Wark, 5 November 1813, B. H. Latrobe MS. Coll., MHS.

[19] B. H. Latrobe to Robert Fulton, 15 March 1814, B. H. Latrobe MS. Coll., MHS; B. H Latrobe to Nicholas Baker, 27 January 1814, B. H. Latrobe MS. Coll., MHS.

[20] *Pittsburgh Gazette,* 9 April 1813.

[21] Ibid.

[22] Ibid., 7 May 1813.

[23] B. H. Latrobe to Robert Fulton, 15 March 1814, B. H. Latrobe MS. Coll., MHS.

[24] Copy of letter, Thomas Copeland to B. H. Latrobe, no date, enclosed in B. H. Latrobe to Robert Fulton, 3 June 1814, B. H. Latrobe MS. Coll., MHS.

[25] *Niles' Weekly Register,* volume 8 (29 April 1815), page 141.

[26] B. H. Latrobe to Nicholas Roose-

velt, 1 August 1815, B. H. Latrobe MS. Coll., MHS.

[27] *Pittsburgh Gazette,* 17 July 1812; Bathe, *Evans,* page 218; Solon J. and Elizabeth Hawthorn Buck, *The Planting of Civilization in Western Pennsylvania* (Pittsburgh, 1939), page 316.

[28] *Pittsburgh Gazette,* 5 February 1819; B. H. Latrobe to William Wilkins, 13 August 1816, B. H. Latrobe MS. Coll., MHS.

[29] Jones, *Pittsburgh in . . . 1826,* pages 60-61; W. G. Lyford, *The Western Address Directory . . . (For the Year 1837)* . . . (Baltimore, 1837), page 97.

[30] McLane Report, volume 2, pages 251-252.

[31] *Pittsburgh Gazette,* 5 March 1821.

[32] Jones, *Pittsburgh in . . . 1826,* pages 60, 56-57, 61.

[33] B. H. Latrobe to Robert Fulton, 1 March 1814, B. H. Latrobe MS. Coll., MHS; Swank, *Iron in All Ages,* pages 227-228.

[34] Seelye A. Willson, "The Growth of Pittsburgh Iron and Steel," *Magazine of Western History,* volume 2 (October 1885), page 543.

[35] McLane Report, volume 2, pages 239-240; Jones, *Pittsburgh in . . . 1826,* page 61.

[36] B. H. Latrobe to William Stackpole, 6 February 1815, B. H. Latrobe MS. Coll., MHS.

[37] *Pittsburgh Gazette,* 5 February 1819.

[38] Advertisement dated 14 September 1818, ibid., 5 February 1819.

[39] McLane Report, volume 2, pages 303-304.

[40] Ibid., volume 2, pages 252-253.

[41] Richard Claiborne to Oliver Evans, 24 August 1807, Evans papers, Franklin Institute.

[42] H[enrico] M'Murtrie, *Sketches of Louisville And Its Environs . . .* (Louisville, 1819), page 131.

[43] B. H. Latrobe to Samuel Hazelhurst, 31 March 1813, B. H. Latrobe MS. Coll., MHS.

[44] David Prentice to father, 18 March 1813, from the collection of The Filson Club, Louisville, Kentucky.

[45] B. H. Latrobe to David Prentice, 24 March 1813, B. H. Latrobe MS. Coll., MHS.

[46] B. H. Latrobe to Jacob Perkins, 24 March 1813, B. H. Latrobe MS. Coll., MHS. Prentice may have been the man to whom Evans referred in 1812. Speaking of an engine then set up in Pittsburgh, Evans wrote that "this engine was first set up in Philadelphia, to saw timber, by a steam engineer, instructed in the old principles; but he could not make it work above one minute at a time, nor would he suffer me to correct his errors, but took it down and sold it, and set up one on the old principles, which he thought he could make work better . . . ." Evans, "Account," *Archives of Useful Knowledge,* volume 2 (1812), page 364.

[47] B. H. Latrobe to David Prentice, 31 March 1813, B. H. Latrobe MS. Coll., MHS.

[48] David Prentice to father, 18 March 1813, Filson Club.

[49] Ibid.

[50] Thomas Woodhouse Bakewell, sketch of the life of Thomas Woodhouse Bakewell written by himself. 1873, Filson Club. In 1824 Prentice summed up his experience by saying: "I am a millwright, engineer, and steamboat owner; and . . . for two years I ran a steamboat on the Mississippi." David Prentice to Secretary of War, 10 July 1824, in *Henry M. Shreve–Snag Boat,* 28 February 1843,

House Document 272, 27th Congress, 3d session, page 6.

[51] Bakewell, Sketch of . . . , Filson Club.

[52] Ibid. In 1820 Prentice and Bakewell listed their capital investment "in premises & material" at $10,000. Return of Prentice & Bakewell, no. 121 (Kentucky), Schedules of the 1820 Census of Manufactures, Record Group 29, National Archives.

[53] M'Murtrie, *Sketches of Louisville*, pages 131-133.

[54] See obituary of Bakewell in *Scientific American*, volume 30, new series (16 May 1874), page 310; Frederick Brent Read, editor, *Up the Heights of Fame and Fortune, and The Routes Taken by the Climbers to Become Men of Mark* (Cincinnati, 1873), pages 77-78.

[55] Woodbury Report, pages 321-326.

[56] Daniel Drake, *Natural and Statistical View, or Picture of Cincinnati and the Miami Country* (Cincinnati, 1815), page 143.

[57] *Niles' Weekly Register*, volume 15 (10 October 1818), page 112.

[58] B. Drake and E. D. Mansfield, *Cincinnati in 1826* (Cincinnati, 1827), page 64.

[59] Ibid., page 61.

Chapter 6

The Uses of Steam

Rochester [N.Y.] is one of the most famous of the cities built on the Jack and Bean-stalk principle. There are many splendid edifaces in wood; and certainly more houses, warehouses, factories, and steam-engines than were ever collected together in the same space of time; but I was told by a fellow-traveller that the stumps of the forest are still to be found firmly rooted in the cellars.

FRANCES TROLLOPE (1831)[1]

By 1838 the role of the stationary steam engine in the United States was largely stabilized. The types of engines best suited for steamboats and locomotives were known and applied. Steam was used in every conceivable type of manufacture from printing presses to gunpowder mills. The gradual exhaustion of new waterpower sites in the East, the movement of manufacturing to the West, and the outgrowing of local water resources by individual mills, all had led to the general acceptance of stationary steam engines. They were found in every part of the country. The question was no longer whether they would work, but whether they were the most satisfactory solution to a particular problem.

The effect on American manufactures was profound. Once scattered by the geographical imperatives of waterpower, industry, now relatively free, was inclined to concentrate in towns. Mills, once limited in size by the waterpower of their streams, were now as large as the engines that ran them. Of the 3,010 steam engines reported in operation in the United States in 1838, over half were stationary.[2] New England accounted for 317 of these engines. The Middle Atlantic States had 502 and the South Atlantic 295. There were 115 in the north-central section of the country and 370 in the south-central section. By states, the leaders were Pennsylvania with 383, Louisiana with 274, Massachusetts with 165, and Virginia (including the present state of West Virginia) with 124.

Perhaps the most rural site of steam engine use was in the Deep South. Engines had been used there since the turn of the century, and by 1838 Louisiana alone had

STEAM ENGINES[3] IN THE UNITED STATES–1838

State	*Stationary*	*Steamboat*	*Locomotive*
Maine	41	8	2
New Hampshire	6	1	–
Vermont	–	4	6
Massachusetts	165	12	37
Rhode Island	58	2	–
Connecticut	47	19	6
New York	87	140	28
New Jersey	32	21	32
Pennsylvania	383	134	96
Delaware	11	3	14
Maryland	56	19	31
District of Columbia	13	5	–
Virginia	124	16	34
North Carolina	20	11	5
South Carolina	40	22	27
Georgia	23	29	3
Florida	8	17	2
Alabama	40	18	1
Louisiana	274	30	10
Arkansas			
Mississippi			
Tennessee			
Kentucky (and part of Indiana)	–	41	2
Missouri and Illinois	56	42	–
Ohio	83	79	1
Michigan and Wisconsin (in part)	32	13	6
Iowa			
U.S. Government	17	14	–
Total ascertained	1,616	700	337
Add, as estimated	244	100	13
Aggregate	1,860	800	350

more engines than any other state except Pennsylvania. Their main use was in the various extractive industries. As was true in the cities, engines in the South aided but did not seem to create industries. They were generally applied to predominant manufactures: salt in the Kanawah country of Virginia, rice in Georgia, sugar in Louisiana, and sawmills and cotton gins scattered over the whole section. More often than in any other part of the nation, these engines were sometimes of English manufacture.

Unlike the engines found in Philadelphia, for example, which were overwhelmingly of local manufacture, those in the South were imported from other parts of

STATIONARY STEAM ENGINES[4] IN THE UNITED STATES—1838

State	*Number*	*Total h.p.*	*Average h.p.*
Maine	41	765	19
New Hampshire	6	102	17
Massachusetts	165	2,244	14
Connecticut	47	315	7
Rhode Island	58	1,430	25
New York	87	1,415	16
New Jersey	32	516	16
Pennsylvania	383	7,448	19
Delaware	11	88	8
Maryland	56	683	12
District of Columbia	13	206	16
Virginia	124	1,567	13
North Carolina	20	751	38
South Carolina	40	675	17
Georgia	23	799	35
Florida	8	215	27
Alabama	40	800	20
Louisiana	274	7,796	29
Missouri and Illinois (in part)	56	1,120	20
Ohio	83	1,786	22
Michigan	32	368	12
U.S. Government	17	340	20
Total	1,616	31,439	
Add (estimated)	244	4,880	
Total	1,860	36,319	

the country and even from abroad. Engines, like any other type of manufactured product, followed the prevailing routes of commerce. Of stationary engines listed for North Carolina in 1838, thirteen were built in Baltimore, six in New York, one in Philadelphia, and two were unknown.[5] In Florida, four were built in New York, one in Boston, two in Baltimore, and one was unknown. Georgia shows the same pattern: thirteen of the engines were made in New York, three in England by Boulton and Watt, and one each in New Jersey, Baltimore, and Philadelphia. The makers of eight engines were unknown. Louisiana saw the most striking development of this pattern. Of the engines enumerated for that state, thirty-five were built in Cincinnati, sixteen in New York, seven each in England and Louisville, only four in New Orleans, two in Pittsburgh, and one in Philadelphia. These figures show clearly the three major trade routes leading to that state: down the Mississippi from the Old Northwest, down the coast and around Florida from New York, and directly from England.

Sugar mills located along the Atlantic coast, in Georgia and Florida, were generally modest, built to service the small local plantations. In these mills, animal power continued to be relied upon until after the middle of the century.[6] It was estimated that "on high lands, where persons have pasturage in abundance, and do not contemplate going beyond one hundred and fifty acres of land, . . . a cattle mill [would be] quite sufficient—the cost of which, with kettles and every thing else that may be required, at New York, will be, from 1,500 to 2,000 dollars." It was recommended, however, that on larger plantations, such as those in Louisiana, "animals as the moving power of these mills be replaced by the steam engine."[7]

Plantation owners in Louisiana began to invest heavily in sugar-mill engines in the late 1820s. It has been estimated that three were purchased in 1827-28, fifteen in 1828-29, and thirty-seven in 1829-30.[8] In 1838 there were between 130 and 150 engines, and by 1844, there were 408.[9] It was claimed in 1833 that although more than half of the sugar plantations still depended upon cattle for power, well over half of the sugar produced was processed by steam, since the engines were found most often on the larger plantations. At that time, no windmills or water mills were thought to be operating in the state.[10]

The reliance on steam continued to grow. In 1846, of the 762 sugar plantations in the state, 408 reportedly were using steam and 354 still were using animals to operate their mills. A mill and engine cost between five and six thousand dollars, and a total of over two million dollars was invested in this type of equipment. The number of engines was rising rapidly, and at that time "upwards of 70 new engines and sugar-mills . . . [were] under contract in Cincinnati alone, . . . and no doubt a still larger number at Pittsburgh, Philadelphia, and New York."[11] By 1861 the transformation was almost complete. Of the 1,291 sugarhouses in Louisiana, 1,027 used steam. At the same time in Texas only three or four mills still used animal power.[12]

Rice mills were to South Carolina what sugar mills were to Louisiana. Throughout most of the eighteenth century, South Carolina planters cleaned their rice with hand mortars and pestles. Water-powered mills were introduced during the 1780s, and in 1791-92 a member of the enterprising Lucas family is supposed to have constructed the first tide-mill to clean rice. In 1838 twelve of the forty-two engines enumerated for South Carolina and eleven of the twenty-four in Georgia were used in rice mills. These mills were concentrated in Charleston and Savannah. In 1846 it was stated that "at present almost every planter of four hundred acres and upwards, is provided with a tide-water or steam-pounding mill for preparing his own crops for market." The smaller planters used custom mills, grinding for toll, that were located in the larger towns. These were usually steam driven. Some rice was shipped unprocessed to the North, to be worked up in large, steam-powered custom mills in New York and Boston.[13]

One of the earliest engines to be used in a rice mill was that bought in 1815 from Boulton and Watt by McAlpine and McInnis for their mill in Savannah. This was a 31-inch cylinder, 90-horsepower engine, which was still operating in 1894, supplying power for the Brush Electric Light and Power Company of Savannah.[14]

Other industries, tied to the agricultural productions of the South, also utilized steam power. William Longstreet operated steam engines to run cotton gins of his own invention in Georgia during the War of 1812. When the steamboat *Comet,* built by Daniel French in 1812-13, was broken up in 1814, its engine was taken ashore and put to use in a cotton gin.[15] Fairly large cottonseed oil factories were built at a later date. One put up in Natchez in 1833 contained an engine which was used for "driving eight hulling machines, five sets of stones, and a machine to prepare and grind the seed for heating."[16] Engines were also purchased by planters for general farm purposes. An engine could be had as cheaply as a slave and could do much more work. Before the Civil War, a firm in Charleston, South Carolina, was specializing in small, portable engines for planters: three-horsepower engines that could be pulled by a single horse. These engines were reported to be "made of the best materials, and . . . very useful for planters."[17]

Southern capitalists were more conservative about building cotton textile mills, but when they were decided upon, steam power was frequently adopted to operate them. A few steam mills began to appear in the 1830s, one of the earliest being Mount Hecla Mills at Greensboro, North Carolina. It originally had been built in 1818, but twenty years later an engine from Paterson, New Jersey, was installed. Five more steam cotton mills were built in North Carolina between 1836 and 1862.

The drive to introduce cotton mills in the South accelerated during the 1850s, and reports of their openings and operations filled the contemporary press. A 50-horsepower engine was installed in the Mobile Cotton Factory in 1850.[18] Another cotton factory in Mississippi operated with a large engine made by Thurston, Green, and Company of Providence, Rhode Island. This engine was tended "by a negro engineer, who also serves as fireman, who had no acquaintance with engines until he took hold of this."[19] A new mill was reported in Memphis, run by an engine from New Albany, Indiana.[20] In 1852 Georgia had thirty-six cotton mills, thirty-four of which were operated by water and two by steam.[21] By 1850 there was a steam cotton mill in Charleston, South Carolina, "worked entirely by white labor."[22]

This Charleston mill was built by Charles T. James (1804-1862), of Rhode Island, one of the most active of the consulting engineers and machine manufacturers of the immediate antebellum period. James had been educated as a carpenter and while working in a machine shop had made himself familiar with all the steps in the manufacture of cotton machinery. He became superintendent of Samuel Slater's steam cotton mill in Providence, but then established his own business of erecting cotton mills. James built mills in Massachusetts, New York, Pennsylvania, Indiana, Tennessee, and South Carolina. It was declared in 1850 that "Mr. James, when his present contracts shall have been completed, will have put in operation a number of spindles considerably exceeding 300,000, being more than one-eighth of the entire number in the United States."[23] His advocacy of steam power, and Southern manufactures generally, won for him the respect and acclaim of such Southern nationalists as J.D.B. De Bow.

The engineering services offered by James were comprehensive. His engines and other machinery were all of his own design, and "every engine and boiler, all the

Charles T. James of Rhode Island.
From *DeBow's Southern and Western Review,* volume 9 (December 1850), facing 577.

shafting and machinery, together with the building, foundations, and every thing else connected with the mills, are designed and drawn in his office, either by himself, or under his immediate direction. Of all these, duplicates are furnished to the workmen in the several departments, and the originals kept on file." As a result of this detailed planning, "his mills are different in their general arrangement, and in many of their details, from any others in the United States. A scientific adaptation of every machine to every other machine, causing the whole to work in unison, without excess or deficiency in any one, produces harmony, and a perfect effect."[24] A few of the other large machine-making firms in the North furnished services similar to those offered by James.[25]

Like their northern counterparts, southern cities also found steam a useful stimulus to local enterprise. The 1848 census of Charleston, South Carolina, counted fifty-nine manufacturing establishments either within the city itself or in Charleston Neck, outside the corporate limits. These appear mostly to have been small establishments, set up to process raw material raised in the hinterland. The largest number (fourteen) were "Grit Mills," five handled cotton, six processed rice, and five were saw mills. Of these fifty-nine manufactories, twenty-eight used steam power, seventeen were powered by animals (including all twelve of the gritmills), twelve operated on manual labor (notably the shipyards, carriage factories, stone masons, and the brass and copper foundry), and one sawmill used waterpower. Five of the establishments listed were called "Steam Engine Manufactories and Casting Establishments." These not only made engines but also other sundry machinery.[26]

The aggregate number of engines in Charleston may not have been large in absolute terms, but it represented a real step forward for the mechanic arts in that city. Only a few years earlier the textile manufacturer and industrial propagandist William Gregg had heaped scorn on the city fathers for forbidding the use of steam within the city limits. He complained, "The labor of negroes and blind horses can never supply the place of *steam,* and this power is withheld lest the smoke of an engine should disturb the delicate nerves of an agriculturalist; or the noise of the mechanic's hammer should break in upon the slumber of a real estate holder, or importing merchant, while he is indulging in fanciful dreams, or building on paper, *the Queen City of the South*–the *paragon* of the age."[27]

The sensibilities of southern gentlemen in Savannah may have been as delicate as those of the Charleston gentry, but they were less naked in their enmity toward the steam engine. The compiler of a Savannah census in 1848 was able to boast that "her growing population–the great increase of the mechanical arts–the extended use of steam as applied to mills, presses, and other useful employment–are all evidences of a healthful state of the body politic." Possibly the gentlemen of Savannah were merely more careless than their brothers in Charleston. The census taker added, "this increase of steam power has been so noiseless, that it has attracted the notice of but few of our citizens, and most of them will be surprised to learn that of the *eighteen* establishments propelled by steam, *fourteen* of them have been erected within the last ten years."[28]

The population had grown from 5,195 to 13,573 between 1810 and 1848, and a large number of the employed male adults were connected with the mechanic arts in some way. Out of a total of 2,201 listed as employed, 380 were given as mechanics, 1 as a civil engineer, 30 as just engineer, and 13 were connected with steam mills and cotton presses. Even among the Negroes, there were 34 free and 73 slave mechanics, and 2 free and 4 slave engineers and pilots. Of eighteen steam mills counted, four were cotton presses, three were rice mills, seven were saw mills, three made blinds, sashes, and other wooden fixtures, and two were iron foundries.

The two largest establishments in town were the Upper Steam Rice Mills and the Vale Royal Steam Saw and Planing Mill. The former was described as being "on the Savannah river, in Ogelthorpe ward–has three engines–two of forty horse power, and one of twenty horse power. Number of buildings 7, all of brick. Beats out 150 tierces of rice per day. . . . Employs about the mill 48 persons, and four or six months in the year about 50 colored women in addition; consumes 1600 cords of wood per annum." The saw mill,

on the Savannah river, at the West end of the city, near the mouth of the canal . . . commenced in February, 1848, and put in operation July 1, 1848. This mill . . . is a noble structure on an extensive scale, and very complete. The main building is 130 feet by 50 feet. Has three engines of thirty horse power each, running three gangs of saws for sawing lumber, and four circular saws for making shingles and laths and edging boards, and two planeing machines. Can saw out 20,000 feet of long lumber per day.

The mill represented a capital investment of $70,000 and employed fifty persons.[29]

Steam engines were needed as much for the expansion of manufactures in the Old Northwest as they were in the South. Those ample and turbulent streams which in the East defined the locations of great cities by meeting tidewater at the Fall Line were wholly absent in the West. On a tour of the western frontier in 1770, George Washington noted carefully in his diary the location of a "small Water" tributary to the Ohio. He did this, as he explained, "because of the Scarcity of them and to show how badly furnished this Country is with Mill Seats."[30]

Most streams in that section were ill suited to the moving of machinery. Furthermore, the problems of using waterpower, common to all sections of the country, seemed to be accentuated in the West. Mill races froze in the winter, flooded in the spring, dried up in the summer, and were clogged with leaves in the fall. Illinois perhaps suffered more than other areas in this respect. One gazetteer, who might be expected to put the best face possible upon conditions, admitted in 1837 that "in the infancy of a state, little can be expected in machinery and manufactures; and in a region so much deficient in water-power as some parts of Illinois are, still less may be looked for."[31] Resort was made to "ox-mills on the inclined plane, and horse-mills by draught," but these were patently expedients.[32] It was predicted that "Steam Mills will become numerous, particularly in the southern and middle portions of the state."[33]

One of the earliest and most important uses for steam in the West was for grinding flour. The first steam-driven gristmill west of the mountains was that of Oliver Evans, built in Pittsburgh in 1809. The want of such a large merchant mill had long been felt, and Evans reported that "previous to this mill's going into operation, Pittsburgh had been no market for wheat; but afterwards, on the first good sleighing, wheat came from every quarter the distance of forty miles. Forty sleds were at one time waiting to be unloaded, and ten thousand bushels were received in a few days; and the wheat farms soon raised fifty per cent in price." Perhaps part of Evans' reason for building this mill was promotional, since it was his opinion that "inventors must apply and test their inventions themselves, or they, most probably, will remain dormant."[34]

The most spectacular steam gristmill in the West, and perhaps in the country, was the one built at Cincinnati from 1812 to 1814. Oliver Evans' son George supplied the plans and built the 70-horsepower engine to drive it. Located on the banks of the Ohio River, the mill was built of stone and was nine stories high. Through the building there was a "wall dividing each story into two unequal apartments—the one designed for manufacturing flour; the other for receiving wool and cotton machinery, a flax seed oil mill, fulling mill, and several other machines."[35] The two sets of grindstones of the flour-milling section were able to grind 36½ bushels of wheat in one hour.[36] After this great mill burned, it was rebuilt in 1823 and finally was sold at a great loss in 1835.[37] The citizens of St. Louis, Missouri, saw early the importance of building steam flour mills, and in the years before the Civil War this river town became the first in the country to develop into a major milling center without the benefit of any waterpower.

As the usefulness of steam engines was demonstrated, most towns of any pretensions in the Ohio Valley established steam mills. In 1817 Steubenville, Ohio, then founded only 19 years, had "an extensive woolen factory, and a very large paper mill, both worked by steam; an air foundry, a brewery, a steam flour mill, a steam cotton factory, a nail factory, etc. etc."[38] Eight years later, besides the great steam-driven woolen mill, the manufactories of the town included "two steam flour mills—two steam cotton manufactories—one steam paper mill—two breweries—two copperas manufactories—one air foundry, in which are cast all kinds of hollow ware—one steam engine manufactory—one cotton and woollen machinery manufactory—two wool carding machines for country work, one of which is propelled by steam."[39]

Farther north, even Michigan (admitted to the Union in 1837) was using steam extensively by 1838. Here, as in the South, steam was used to encourage the prevailing manufactures of the area. Of the engines listed in 1838, twenty-eight were used in sawmills. Twenty-two of these engines were built in Detroit, which had already become an engine-building center. The rest had come across Lake Erie: eight from engine shops in upper New York State, seven from Cleveland, and one each from Connecticut and Pittsburgh.[40]

Pittsburgh was still the steam engine and manufacturing center of the West. In 1833 there were approximately eighty-nine engines in the city and its vicinity. Five years later there were at least 133. The greatest single use for steam engines in Pittsburgh was in the growing iron industry. These engines were among the largest in the country: horsepower ratings of over one hundred were not uncommon, and in 1829 an engine of 190 horsepower was built for the Kensington Iron Works by Samuel Smith. The manufactures of the city—and the number of steam engines—continued to grow, though ironworks, glass works, and engine shops still dominated industrial activities.[41]

So extensive was the use of steam engines and stone coal in Pittsburgh that this city early received its reputation as a smoky, dirty city. In 1841 a gazetteer described the sight which confronted a traveler as he approached Pittsburgh:

> A dense cloud of darkness and smoke . . . visible for some distance before he reaches it, hides the city from his eyes until he is in its midst; and yet, perhaps half this volume is furnished by household fires, coal being the only fuel of the place. As he enters the manufacturing region, the hissing of steam, the clanking of chains, the

View of the Kanawha Salines, 1835.
From *The American Journal of Science and Arts,* volume 29 (January 1836), page 34.

jarring and grinding of wheels and other machinery, and the glow of melted glass and iron, and burning coal beneath, burst upon his eyes and ears in concentrated force. If he visits the warehouses, he finds glass, cotton yarns, iron, nails, castings, and machinery, occupying a prominent place. He discovers the whole city under the influence of steam and smoke. The surface of the houses and streets are so discolored as to defy the cleansing power of water, and the dwellings are preserved in any degree of neatness, only by the unremitting labors of their tenants, in morning and afternoon ablutions. The very soot partakes of the bituminous character of the coal, and falling—color excepted—like snow-flakes, fastens on the face and neck, with a tenacity which nothing but the united agency of soap, hot water, and the towel can overcome. Coal and the steam-engine are the pervading influence of the place, and over the whole city the seal and impress is—'Great is Vulcan of the Pittsburgers.'[42]

Although wood remained the universal fuel in the nation as a whole—an 1828 fuel estimate gave 80 percent wood, 3 percent charcoal, 14 percent anthracite coal, and 2 percent bituminous—coal was widely used at Cleveland and along the Ohio River. In many river areas coal could be had for one or two cents per bushel. A western correspondent of Silliman's *Journal* wrote in 1836 that

its great abundance and cheapness have given birth to the vast and widely extended manufacturing establishments of the west, there can be no doubt. Without coal they could not exist. It constitutes the life spring of western Pennsylvania, and the pedestal of our great manufacturing emporium. Pittsburgh and its environs contain ninety steam engines for the various manufactures of iron, steel, glass, cotton, salt, brass, white lead, flour, oil, leather, &c. &c. These engines consume two millions sixty five thousand three hundred and six bushels a year.[43]

The Kanawha salines in western Virginia provided an excellent example of the way in which steam engines furthered local industry without being necessary to the introduction of that industry. Salt was a commodity of great value and rarity on the western frontier. The first settlers on the Kanawha River were fortunate to find salt water, not only because of their own needs but also because of the gathering of buffalo around the salt lick, which made it an excellent hunting ground. These early settlers found evidence that the Indians had used the same site for many years. The first salt furnace was put up about 1797 near the present site of Charleston. In 1807 the Ruffner brothers began full-scale operations in the area, with improved equipment and techniques, and the Kanawha salines began to attract a more than local reputation.[44]

By 1814 annual production had risen to 600,000 bushels of salt, and three years later coal was first mined in the area for use as furnace fuel. The earliest method of obtaining the salt water had been to dig a pit and allow the saline to rise and be collected in buckets. An improvement over this technique came with the introduction of pumps operated by horse power. About 1828 steam engines were introduced for pumping and for drilling. A visitor to the area in 1836 remarked that "at this period, a large portion of the furnaces have a small steam engine attached for the purpose of raising water, which contains more salt the nearer they approach the bottoms of the wells." The amount of water to be pumped was considerable since it took 70 gallons of saline to make a bushel of salt, and by 1836 production had risen to more than one and a half million bushels annually. This same observer noted that

> the Kanawha salines present a most interesting and lively scene of activity and business. At intervals of every quarter of a mile, both shores of the river are lined with furnaces, sending forth dense curling volumes of coal smoke. The busy hum of voices, and the rattling of the 'train wagons,' along the rail ways, with the bustle of the salt boats, and steam boats, to which the depth of the river affords a safe and pleasant navigation to the upper furnaces, give this spot all the life and activity of a large city.

He noted also that "mineral oil [petroleum] rises in nearly all the wells," and a later historian of Kanawha County pointed out that "the Kanawha borings have educated and sent forth a set of skilled well-borers, all over the country, who have bored for water for irrigation on the western plains, for artesian wells for city, factory, or private use, for salt water at various places, for oil all over the country, for geological or mineralogical explorations, etc. etc."

That the area did present a "lively scene of activity and business," as the visitor of 1836 observed, is upheld by the statistics on steam engines collected two years later. Sixty-two engines were operating at salt works, two others powered sawmills, and one was at work in a "mill, foundry, & machine shop." Actually, a number of the engines used primarily for pumping did double duty by grinding grain. The oldest in operation had been put up in 1828, a fair number had gone up in 1829, and the rest were set up mainly in the early 1830s. With the exception of one from Cincinnati, the engines were manufactured in Pittsburgh. All were high-pressure engines, and most were quite small, listed at two or three horsepower.

New England was long thought of as the section of the United States most generously endowed with waterpower and most reticent to adopt the motive power of steam for manufactures. It is true that the northeastern section of the country did lag behind the rest of the nation in the adoption of steam engines, and especially in the manufacture of steam engines, but the lag has been exaggerated. The impressive waterpower developments in such textile centers as Lowell, Lawrence, and Fall River have tended to obscure the wide and varied use of steam among smaller and more diverse urban manufactories. In 1838 when the debate over the comparative economies of steam and waterpower was just beginning to reach major proportions, there were 114 steam engines in operation in the greater Boston area. The size of the figure makes it evident that steam was accepted in New England long before the Civil War.

The early use of steam in New England was contemporary with the general rise of manufacturing activity during the War of 1812. One of the first engines in the

STEAM ENGINES IN PITTSBURGH AND VICINITY, 1833-1838[45]

Mills	*1833*	*1838*
Flour	4	6
Saw	7	10
Engine and machine	7	11
Glass	6	77
Tanneries	5	7
Turners (wood and iron)	6	18
Distilleries	3	
Printing presses	1	1
Brass foundries	1	2
White lead	5	6
Fulling and carding	1	
Edge tools	5	3
Rolling iron	8	12
Iron foundries	1	4
Waterworks	2	3
Breweries	3	3
Oil	2	1
Cotton	4	7
Wool		2
Paper	2	2
Nails		5
Wood planning		4
Rope		2
Total Engines	89	133

area was built by Oliver Evans in 1811 for the Middletown Woollen Manufacturing Company of Connecticut. This engine was in operation by May 1811 and drove "all the machinery for carding, spinning, reeling, weaving, washing, fulling, dyeing, shearing, dressing, and finishing."[46] After a year of operation, the English "chief artist" of the mill wrote Evans that the engine was so far superior to those of previous designs, that "I would not take them as a gift, could I obtain yours at your price." Evans considered this high praise from a man who had "been in England 13 years, engaged in manufactories wrought by steam engines."[47] Since Evans' engine did not condense the steam it used, the steam was available for further use after its expansion. In the Middletown mill, it was used to heat the building during cold weather and was also "applied in connection with the brushing machine in finishing their cloth, without adopting the method of oiling and hot pressing as is commonly practiced in England. In this method of finishing," it was explained, "the cloth does not require sponging before it is made up."[48]

In 1812 another Evans engine was sent to the Providence Woolen Company in Rhode Island. Two years later an Evans engine of 24 horsepower was purchased by the Providence Dyeing, Bleaching, and Calendering Company. This engine was reported to have cost seventeen thousand dollars, a large portion of which was spent for transportation overland from Philadelphia.[49]

One of the most publicized of the early New England cotton mills was that of Samuel Slater, begun in Providence in 1827. An engine was purchased, probably from Rush and Muhlenberg (successors to Oliver Evans), and the products of the mill were proudly labeled "Steam Loom."[50] Slater commented in 1832 that "with such a redundance of [waterpower] . . . , the more expensive one of steam has been, so far, very little used, and only under peculiar circumstances of location and business. Four cotton mills," he added, "are now driven by steam, and two more, to be operated by that agent, are in progress. Most of the bleacheries and calenders, and some machine shops, foundries, and other works for iron, make use of the same agent."[51] Slater was considered the father of cotton manufactures in the state, and his building of a steam mill in 1827 no doubt lent confidence to that method of manufacture.

Steam cotton mills were also built in other parts of Rhode Island. Perhaps the first was erected by Oziel Wilkinson in 1810-11 at Pawtucket. David Wilkinson, the former partner of Elijah Ormsbee, helped build the machinery for this pioneer mill.[52] By 1835 there were two steam cotton mills operating in Newport, Rhode Island, with another under construction. At this time it was pointed out that "it is cheaper to use steam power in the midst of a dense population, than to use water power, which often makes it necessary not only to build a factory but a town also."[53]

Providence, however, remained the chief center of steam power in the state. Twenty-three engines were at work in the city in 1838 and by 1850 this total had grown to sixty-five.[54] In 1856 the statement was made that "the whole number of stationary steam-engines in operation within the limits of the city is 73. There are also 12 to 15 more within 100 rods of the city, which are not included."[55] The size as well as the number of engines was becoming larger. Samuel Slater had estimated in 1832 that all of the engines in Rhode Island (including those used for steamboats) did not develop more than 800 horsepower.[56] One engine alone in 1856 developed 400 horsepower, and the total normal capacity of the engines in Providence amounted to 4,332 horsepower, or an average of nearly 60 per engine.[57]

In Massachusetts, Boston and its immediate vicinity had twice as many engines in operation in 1838 as the rest of the state combined. Daniel Treadwell, who began to manufacture his printing presses in that city about 1821, later declared that "there was not at that time . . . a single steam-engine at work in any shop or manufactory in the old peninsula of Boston, and but a single one at the foundry at South Boston."[58] Actually, Treadwell may have overlooked a few engines here and there about the city. One of Samuel Morey's rotary engines, for example, was operating a sawmill in the Charlestown Navy Yard as early as 1818.[59]

By 1838 the number of engines in Boston had jumped to 114, or two-thirds the number for the entire state of Massachusetts. Sixteen, the largest number, were used in sawmills and woodworking shops; fifteen in machine shops; thirteen in ropewalks and cordage factories; eleven in various types of ironworks; six in tanneries; and six in distilleries and breweries. The other engines were scattered through different kinds of shops—chemical laboratories, laundries, oil mills, an India rubber manufactory, and a bone mill[60]—demonstrating the diverse manufactures of the city and the wide extent to which steam was accepted as a motive power.

In 1838 the relatively small number of steam engines in New England, outside of the large metropolitan center of Boston, was because of the great abundance of waterpower and the lack of any local source of fuel for steam furnaces. The abundance of waterpower in the countryside is axiomatic to any economic study of New England. While figures do not exist for the earlier period, in 1870 water in New England supplied 99,073 horsepower for cotton mills alone, and by 1905 that figure had increased to 251,884.[61] The increase in steam power during these same years was even larger, from 46,967 in 1870 to 702,023 in 1905, but it is significant that—through the development of marginal power sites, improvements in dam engineering, and the use of more efficient turbines—total horsepower from water was still increasing into the twentieth century. With such water resources available, the need for steam power was not felt as strongly or as soon here as in other parts of the country.

The shortage of adequate fuel was also a retarding influence on the adoption of steam in New England and elsewhere. Wood was in such short supply that coal seems to have been used in northern engines almost from the beginning. The use of anthracite coal began in earnest in 1825 in the general vicinity of Philadelphia. Three years later, 82,302 tons of coal reached Philadelphia by way of the new coal canals. Of this total, 19,000 tons were forwarded to New York, some of which, no doubt, was sent on to New England.[62] Samuel Slater used Pennsylvania anthracite at his Providence cotton mill as early as 1830. He experimented with both Lehigh and Lackawanna coal, the latter purchased from the Delaware and Hudson Canal Company in 1831 at six dollars per ton.[63] Shipments of coal to Boston rose rapidly after 1825, and by 1841 that city received 110,432 tons of anthracite in addition to 124,041 bushels of bituminous from Richmond, Virginia, and 12,754 tons of foreign coal.[64]

The mines near Richmond were extensive and provided the bulk of domestic American coal until the anthracite region of eastern Pennsylvania went into production. A survey in 1818 found at least twenty-five different pits opened, some of which had been worked for thirty years. One local operator, perhaps working on a larger scale than most, had reopened an old mine abandoned some years before when inefficient methods of extraction proved unprofitable. The new owner imported two miners from Scotland and a Boulton and Watt steam engine. Located at the deepest (300 feet) shaft, the engine pumped out water in the manner "of the same operation at the Cornwall mines."[65] It was the shutting off of this coal by the British blockade during the War of 1812 that so disrupted normal ironworking activities in Philadelphia and elsewhere.

In 1846 the Fall River (Massachusetts) Iron-Works Company showed more ingenuity than most New England firms in ensuring a fuel supply by purchasing its own coal mine near Cumberland, Maryland. This ironworks operated five steam engines as well as various furnaces and reportedly consumed 17,050 tons of coal yearly. Its new coal mine had rail connection with the Baltimore and Ohio Railroad, and the purchase led one observer to comment that "Massachusetts is thus destined to share in the benefit of the Cumberland mines, by making them tributary to her own industry."[66]

The upstream abundance of waterpower and shortage of fuel combined to assure that most of the steam engines in New England would be located in the coastal cities, where waterpower was scarce and transportation costs of fuel at a minimum. For example, of the forty-one engines operating in Maine in 1838, at least thirty-six were located along the coast, mostly in sawmills and ironworks.[67] As the nineteenth century progressed, many of the seacoast towns of New England—formerly commercial ports of some importance—suffered from mercantile competition with New York City. The location in these towns of both idle capital and an idle "female population" is supposed to have entered into the calculations of those capitalists who built large steam cotton mills there.[68]

Despite talk of "purifying the public morals" in crowded seaport towns by "furnishing profitable employment to the thousands of the industrious poor," it was always easier to build a mill in a town than a town at a mill.[69] As one observer pointed out, "a water-mill is necessarily located in the country afar from the cities, the markets and magazines of labor."[70] Even so, the rapidly expanding transportation system of the country was acting to equalize the factor of location. It was estimated that "railroads tend to lessen the difficulties of transit with inland locations, in about the same ratio as the reduction in expense of motive power by the great improvements in the steam engine." There was no necessity, it was alleged, "to array Water and Steam Power as antagonists, both . . . can be used to advantage under the judicious and skillful management of the scientific American manufacturer."[71] The controversy over motive power, which attracted so much attention during the 1840s, was confined largely to the location and economy of large textile mills. For the generality of manufactures, however, there was little debate over the relative merits of water and steam. When water was already being used, it continued to be used; where water was unavailable, steam was seized upon as the obvious alternative.

The steam engine was largely an urban phenomenon. For example, in Pennsylvania in 1838, the Pittsburgh area accounted for 133 engines and Philadelphia for 174–307 out of a total for the state of only 383. In the same year, out of a total of 23 engines in Georgia, 19 were in Savannah; of the 165 engines in Massachusetts, 114 were found in and around Boston.[72]

Philadelphia was typical in its use of steam engines. The 174 engines in that city were manufactured by forty-four different individuals or firms. This number is considerably larger than the number of enginemakers active in the area. It was not unusual for a mechanic of universal talents to make his own engines and then to use them for some other kind of manufacture. Samuel Jackson, of Jackson's Turning & Cotton Machine Manufactory, purchased a high-pressure, horizontal engine from Prosper Martin in 1832. In 1838 he built a larger high-pressure, horizontal engine. It was the only engine in town listed as having been made by him.[73]

Engines in Philadelphia were used not so much to introduce new types of industry as to encourage those for which the city was already famous. More engines were used in machine shops than in any other single type of manufactory. Twenty-seven were used in machine making, and another sixteen were applied to other branches of the iron business. Although their use tended to concentrate in the already predominant industries, engines were applied to a wide variety of other purposes. Well over twenty-five types of mills were using them in the city in 1838,

USES OF STEAM ENGINES IN PHILADELPHIA–1838[74]

Engine and machine making	27
Bleaching, dyeing, and printing cloth	19
Iron	16
Woolen mills	11
White lead	11
Sawmills	10
Printing presses	10
Cotton mills	9
Breweries and distilleries	8
Paper	5
Brass foundries	5
Bark mills	5
Drug mills	4
Flour mills	4
Sugar refineries	3
Sawing stone	3
Tool making	3
Carpet weaving	3

ranging from shops for the manufacture of surgical instruments to stone yards where marble was sawed. In 1831 there were from 90 to 100 engines in the city. This number had risen to 174 seven years later, and by 1853 it was remarked that "so much are manufactories the order of the day, that it is said in no community of the same size on the globe, are there so many steam engines at work, nor any in which they are applied to so great a variety of purpose as in Philadelphia."[75]

After the War of 1812, Philadelphia, long the commercial center of the country, began to lose that position to New York City. Also, much of its inland trade was threatened by the enterprise of Baltimore. A partial solution to this problem could be found in the encouragement of manufactures within the city. The British consul at Philadelphia noticed that although New York excelled in the improvement of trade, Philadelphia had a "manifest superiority as regards manufactures."[76] Any expansion of manufactures, however, was handicapped by a lack of sufficient waterpower. The city was located on the point between the Delaware and Schuylkill Rivers, but neither had sufficient fall within the city to provide much power. Furthermore, both rivers were so large as to discourage any attempt to develop them. It was not until 1825 that the Schuylkill Navigation Company provided waterpower as a side product of navigational improvement, and then—in lieu of cash salary[77]—the company gave a monopoly on it to its engineer, Josiah White.

As early as 1787 Tenche Coxe, addressing a Philadelphia group sympathetic to domestic manufactures, noted that although "steam mills have not yet been adopted in America . . . we shall probably see them after a short time in New-England and other places, where there are few mill seats and in this and other great towns of the United States." According to his calculations, "Philadelphia, by adopting the use of them, might make a saving of above five *per cent.* on all the grain brought hither by water, which is afterwards manufactured into meal, and they might be usefully applied to many other valuable purposes."[78] Coxe's analysis of the future role of steam was essentially correct, even though he reckoned without the future development of large-scale hydraulic engineering projects, such as the damming of the Schuylkill, Merrimack, and other rivers of potentially large but as yet unharnessed power.

Another solution to the power problem of Philadelphia was suggested as early as 1814. In that year Thomas Cooper noted that although "no fall is gained from the Schuylkill to the Delaware . . . there are always machines that do actually raise the water for the purpose of renting it out in portions." It was his suggestion that steam engines be used to raise water in this manner and that individuals rent a portion of "small streams falling from a considerable height." In this way, "a man might afford to pay a high rent for the hundredth or five-hundredth part of the power of a twenty-horse engine, who could not afford to erect such an engine himself."[79]

A more practical solution resorted to in many of the larger cities of the country was that of renting power from a shaft or belt. A promotor in Cleveland, Ohio, in 1854, advertised "MECHANICS' ROOMS, WITH STEAM POWER, to Rent." He had built, especially for this purpose, a three-story brick building, one hundred by two hundred feet. This building was provided with "two large steam engines expressly designed for the growing mechanical wants of this vicinity." Interior space and the power of the engines were to be subdivided according to the needs of the tenants, who could rent any fraction desired. The advertisement concluded that "the location is central, conspicuous, and convenient to canal, railroads and lake shipping."[80] Similar facilities were available in New York City as early as 1848.[81] While the increasing use of steam engines made possible ever larger manufacturing establishments, it also provided for the perpetuation of small, independent operations, using a fraction of an engine where a fraction of a millsite was unavailable.

To further encourage this movement, it was suggested in 1826 that the city of Philadelphia might, "for a limited time, . . . allow the free use of the Schuylkill water, for the supply of all the steam engines, which should be built, and applied to manufacturing purposes, within a certain period." It was thought that precedent existed in the practice of allowing free water for fire fighting—a practice that, like the encouragement of steam engines, was productive of the common good.[82] Apparently no action was taken on this suggestion.

While steam engines made possible an urban concentration of manufactures they also, for the time being at least, helped some mills remain in the countryside. An engine could be purchased and installed at a water mill to act as an emergency source of power during periods of drought or ice, or it could be used to provide additional power which might be required by expansion of business activities. The remark was made that at many waterpower sites around Philadelphia, "it has been more convenient to add steam, where they are than to change the location of the factories."[83] Commenting that "the perfection of combination cannot be attained except by aggregation in a suitable locality," one Philadelphia observer exhorted mill owners to "come out from sylvan retreats, [and] deny themselves the advantages of mill-races and the harmonies of frog-ponds."[84] Some small country mills did heed the call of the large manufacturing centers; others stayed by the frog ponds and denied themselves the benefits of combination and aggregation.

By the middle of the 1830s, stationary steam engines were no longer novelties in any part of the country. They were operating in every state of the Union and were applied to every conceivable purpose. They caused no great revolution in the types of manufactures carried on in any section, but were quietly substituted for less satisfactory motive powers wherever possible. Except in the South, where they were applied extensively to the extractive industries of an agricultural economy, steam engines found their greatest usefulness in the cities of the nation, where increasing populations and limited power created a large demand for them.

Despite the continued importance of waterpower, it cannot be said that the general acceptance of steam power was delayed until after the Civil War. The steam engine was a familiar object of the Jacksonian period and had so far gained recognition that Zachariah Allen could write in 1829: "so necessary to the wants of life at the present day are the products of machinery, that the Steam Engine has become almost as important as the plough."[85]

Notes

[1] Frances Trollope, *Domestic Manners of the Americans,* edited by Donald Smalley (New York, 1949), page 376.

[2] Woodbury Report, page 376.

[3] Ibid. It should be noted that these figures represent minimum rather than maximum numbers, since the Report was incomplete, due to the failure of some inspectors to make returns. Engines enumerated in the Report sometimes total more than the figure shown in the table. For this reason figures cited in the text may not always agree with these totals. An interesting analysis of the Woodbury Report is given in Peter Temin, "Steam and Waterpower in the Early Nineteenth Century," *Journal of Economic History,* volume 26 (June 1966), pages 187-205.

[4] Woodbury Report, page 379.

[5] For details on types and prices of engines suitable for plantation use, built by Fawcett and Preston, of Liverpool, see [Benjamin Silliman], *Manual on the Cultivation of the Sugar Cane, and the Fabrication and Refinement of Sugar* (Washington, D.C., 1833), page 47. The two unknown engines listed may have been built in Leeds, England. Discrepancies in total number of engines for North Carolina and other states are part of the weakness of the Woodbury Report and cannot be resolved.

[6] J. Carlyle Sitterson, *Sugar Country: The Cane Sugar Industry in the South, 1753-1950* (Lexington, 1953), page 138.

[7] [Silliman], *Sugar Cane* page 46.

[8] [Silliman], *Sugar Cane,* page 18.

[9] *Niles' Weekly Register,* volume 41 (September 1831-March 1832), addendum, page 50; J. D. B. De Bow, "Louisiana and Her Industry," [De Bow's] *Commercial Review,* volume 8 (January 1850), page 36.

[10] [Silliman], *Sugar Cane,* page 31.

[11] *Hunt's Merchants' Magazine,* volume 14 (February 1846), page 148.

[12] Sitterson, *Sugar Country,* page 138.

[13] R.F.W. Allston, "Rice," [De Bow's] *Commercial Review,* volume 1 (April 1846), pages 344, 346, 347; *Niles' Weekly Register,* volume 44 (1 June 1833), page 222; *The Mechanic,* volume 3 (March 1834), page 105.

[14] "The Oldest Steam Engine in the United States," *Engineering News,* volume 30 (9 November 1893), page 370; F.R. Hutton, "First Stationary Steam Engines in America," *Transactions of the American Society of Mechanical Engineers,* volume 15 (1894), page 987.

[15] George H. Thurston, *Pittsburgh As It Is; or, Facts and Figures, Exhibiting the Past and Present of Pittsburgh, Its Advantages, Resources, Manufactures, and Commerce* (Pittsburgh, 1857), page 71.

[16] *Niles' Weekly Register,* volume 44 (1 June 1833), page 222.

[17] *Scientific American,* volume 9 (18 February 1854), page 184. For this general topic, see Reynold M. Wik, *Steam Power on the American Farm* (Philadelphia, 1953).

[18] *De Bow's Southern and Western Review,* volume 9 (October 1850), pages 431-432.

[19] Ibid., page 433.

[20] "Memphis, and Its Manufacturing Advantages," ibid., volume 10 (May 1851), page 527.

[21] *Hunt's Merchants' Magazine,* volume 26 (February 1852), page 158.

[22]"Manufactures in South Carolina," [De Bow's] *Commercial Review,* volume 8 (January 1850), page 29.

[23]"Charles T. James," *Appletons' Cyclopaedia of American Biography,* volume 3, page 397; "Charles T. James, of Rhode Island," *De Bow's Southern and Western Review,* volume 9 (December 1850), page 672.

[24]*De Bow's Southern and Western Review,* volume 9 (December 1850), page 672.

[25]See *Scientific American,* volume 5 (27 April 1850), page 253.

[26]J.L. Dawson and H. W. DeSaussure, *Census of the City of Charleston, South Carolina, for the year 1848...* (Charleston 1849), page 172.

[27]William Gregg, "Domestic Industry–Manufactures at the South," [De Bow's] *Commercial Review,* volume 8 (February 1850), page 140.

[28]Joseph Bancroft, *Census of the City of Savannah, together with Statistics Relating to the Trade, Commerce, Mechanical Arts...* (2d edition, Savannah, 1848), page 3.

[29]Ibid., pages 5, 10, 12, 13, 34-36.

[30]John C. Fitzpatrick, ed., *The Diaries of George Washington, 1748-1799* (Boston, 1925), volume 1, page 416.

[31][Samuel Augustus Mitchell], *Illinois in 1837...* (Philadelphia, 1837), page 58.

[32]Ibid., page 59.

[33]J.M. Peck, *A Gazetteer of Illinois...* (Philadelphia, 1837), page 30.

[34][Evans] Elisha, *Patent Right Oppression Exposed,* pages 61-62n.

[35]Drake, *Picture of Cincinnati,* pages 137-138.

[36] *Niles' Weekly Register,* volume 9 (1815-1816), supplement 187.

[37]Ibid., volume 25 (22 November 1823), 192 and *ibid.,* volume 49 (19 September 1835), page 35.

[38]*Niles' Weekly Register,* volume 12 (26 April 1817), page 144.

[39]Ibid., volume 28 (9 April 1825), page 82.

[40]Woodbury Report, pages 346-348.

[41]Ibid., page 193; *Scientific American,* volume 9 (4 February 1854), page 163.

[42]Cist, *Cincinnati in 1841,* page 237.

[43]Howard N. Eavenson, *The Pittsburgh Coal Bed: Its Early History and Development* (New York, 1938), page 3; S.P. Hildreth, "Observations," on the Bituminous Coal Deposits of the Valley of the Ohio. . . ," *American Journal of Science and Arts,* volume 29 (January 1836), page 79.

[44]Material on the Kanawha region is taken from Hildreth, "Observations," *American Journal of Science and Arts,* volume 29 (1836), pages 117-121; James T. Laing, "The Early Development of the Coal Industry in the Western Counties of Virginia, 1800-1865," *West Virginia History,* volume 27 (January 1966), pages 144-155; W.S. Laidley, *History of Charleston and Kanawha County West Virginia and Representative Citizens* (Chicago, [1911]), pages 127, 128; Woodbury Report, pages 226-229.

[45]Figures for 1833 from *Niles' Weekly Register,* volume 45 (30 November 1833), page 217; figures for 1838 from Woodbury Report, pages 191-195.

[46]Evans, "Account," *Archives of Useful Knowledge,* volume 2 (1812), page 364.

[47]*Aurora,* 25 June 1812.

[48]*Niles' Weekly Register,* volume 1 (1 February 1812), page 407. The "chief artist," Isaac Sanford, patented a

"machine for dressing cloth" on 27 March 1799. *Letter from the Secretary of State Accompanied with a list of the names of persons . . . to whom patents have been issued . . .* (Washington, D.C., 1805), page 19.

[49] Hutton, "First Engines," *Transactions of the American Society of Mechanical Engineers,* volume 15 (1894), page 984; *Hunt's Merchants' Magazine,* volume 34 (June 1856), page 673.

[50] Minutes, Providence Iron Foundry, 21 June 1827, volume 1, Slater collection, Baker Library. Harvard University, Cambridge, Massachusetts; H. N. Slater to William Tiffany and Company, 5 March 1830, letter book, volume 14, Steam Cotton Manufacturing Company, Slater collection, Baker Library.

[51] McLane Report, volume 1, page 927.

[52] Edward S. Wilkinson to Elisha Dyer, 16 December 1861, in *Transactions of the Rhode Island Society for the Encouragement of Domestic Industry . . . 1861,* page 88.

[53] *Niles' Weekly Register,* volume 48 (8 August 1835), page 397.

[54] Woodbury Report, pages 88-89; *Scientific American,* volume 5 (20 July 1850), page 346.

[55] *Hunt's Merchants' Magazine,* volume 34 (June 1856), page 673.

[56] McLane Report, volume 1, page 927.

[57] *Hunt's Merchants' Magazine,* volume 34 (June 1856), page 673.

[58] Quoted in Morrill Wyman, "Memoir of Daniel Treadwell," *Memoirs of the American Academy of Arts and Sciences,* volume 11 (1888), page 344.

[59] Deposition of Amos Binney et al., 7 February 1818, in John L. Sullivan, *Explanation, by John L. Sullivan, of the Nature of Certain Grants to Him for the Use of Steam Boats on Connecticut River...* (no place, 1818), page 31.

[60] Woodbury Report, pages 41-44, page 354.

[61] Copeland, *Cotton Manufacturing,* page 28n.

[62] *Niles' Weekly Register,* volume 35 (31 January 1829), page 366.

[63] Zachariah Allen, *Science of Mechanics,* page 352n; H.N. Slater to Major Durfee, 18 July 1830 and Samuel Slater to Delaware and Hudson Canal Company, 23 April 1831, letter book, volume 14, Steam Cotton Manufacturing Company, Slater collection, Baker Library.

[64] *Niles' Weekly Register,* volume 61 (29 January 1842), page 352. In 1840 a New York City editor claimed that he knew of no stationary engine in that city in which fuel other than coal was used. *American Repertory,* volume 1 (March 1840), page 136.

[65] John Grammer, Jr., "Account of the Coal Mines in the Vicinity of Richmond, Virginia," *American Journal of Science,* volume 1 (1818), pages 125-130.

[66] *Hunt's Merchants' Magazine,* volume 15 (August 1846), page 213.

[67] Woodbury Report, pages 20, 27, 30-31, 33.

[68] Samuel Webber, *Manual of Power for Machines, Shafts, and Belts, With the History of Cotton Manufacture in the United States* (Revised edition, New York, 1879), pages 50-51.

[69] John Clowes, "Steam and Water as Motive Powers," *National Magazine and Industrial Record,* volume 1 (October 1845), page 448.

[70] *Scientific American,* volume 4 (12 May 1849), page 269.

[71] Clowes, "Steam and Water,"

National Magazine and Industrial Record, volume 1 (1845), pages 448, 444.

[72]Woodbury Report, passim. It should be noted that engines were more conveniently counted in cities than in rural areas, a fact which no doubt prejudices these figures somewhat. As noted above, Louisiana was an exception to this rule of urban use.

[73]Ibid., page 162.

[74]Ibid., page 156.

[75]*Niles' Weekly Register,* volume 40 (16 July 1831), page 344; Job R. Tyson, "Resources and Progress of Philadelphia," *De Bow's Review,* volume 15 (July 1853), page 48.

[76]Quoted in ibid., pages 376-377.

[77]Taylor, *Transportation Revolution,* page 41; Calhoun, *American Civil Engineer,* page 69.

[78]Tenche Coxe, *An Address to an Assembly of the Friends of American Manufactures...* (Philadelphia, 1787), page 9.

[79]*Emporium of Arts and Sciences,* volume 3, new series (June 1814), page 170.

[80]*Scientific American,* volume 10 (11 November 1854), page 71.

[81]Ibid., volume 4 (4 November 1848), page 55 and ibid., volume 10 (19 May 1855), page 287.

[82]*Franklin Journal,* volume 1 (February 1826), page 106.

[83]*Hunt's Merchants' Magazine,* volume 43 (November 1860), page 558.

[84]Edwin T. Freedley, *Philadelphia and its Manufactures: A Hand-Book Exhibiting the Development, Variety, and Statistics of the Manufacturing Industry of Philadelphia in 1857* (Philadelphia, 1859), pages 46-47.

[85]Allen, *Science of Mechanics,* page iii.

Chapter 7

The Industry Established

In arranging the power for the Newark Machine Works, I procured an engine of about 35 horsepower . . . of good construction. . . . With this engine we drive 14 lathes for iron, 3 for wood work, 4 planers for iron, and one . . . for wood; 4 drills and one boring and one tenoning machine, splinter and bolt cutters, two circular saws, punching and shearing machines, two trip hammers, a large blacking mill, and fan-blower for forges and foundry. . . .

JOSEPH E. HOLMES (1855)[1]

Just as the number and the use of steam engines expanded enormously during the Jacksonian period, so did the making of engines become widespread and systematized. At the end of the War of 1812, enginemakers were few and were concentrated in the three centers of New York, Philadelphia, and Pittsburgh. By the middle of the century, engine shops were found in nearly every sizable town in the country. A general knowledge of steam was even more widely spread. At the turn of the century, the venerable iron manufacturer and mill designer John Fritz reminisced that when be became an apprentice in a Parkesburg, Pennsylvania, blacksmith's shop in 1838, he discovered that his new master had built his own steam engine. Fritz wrote that in those days a mechanic in like circumstances "would have to make his own drawings and patterns, make his own forgings, and fit the work all up, without tools, except makeshifts. To-day, as many men work on an engine as there are parts to it, and each man has a special machine, specially designed to do his work on."[2] The significant point is not that a small-town blacksmith had to improvise in building an engine in 1838, but that he could even build a workable engine.

The Middle Atlantic States remained the center of American engine building throughout the first half of the nineteenth century. Philadelphia had 174 steam engines in 1838, of which at least 133 were built by the forty-four local makers then listed as having engines operating within the city. Rush and Muhlenberg—successors to Oliver Evans—had the largest local trade and were, along with Daniel

Large, the only holdovers from the earlier period. Twenty men were listed as having made only one engine each, indicating that it was not uncommon for a handy mechanic to make an engine on occasion, without having any particular connection with engine building as a business. Among the engine builders mentioned in the Philadelphia city directory for 1840, only one called himself a "steam engine maker," the rest preferring the title "machinist" or "machinist and iron founder."[3]

In addition to its fame as a center of engine and machine building, Philadelphia by the 1850s had also made a reputation in the closely allied manufacture of iron ships. Fabricators of steam boilers, having the necessary equipment and experience in working sheets of iron, pioneered in the building of iron ships in this country. One of the largest firms in this business was Reaney, Neafie, & Company, which was formed in 1844 by two mechanics, Thomas Reaney and Jacob G. Neafie. The latter had been an apprentice of Thomas Halloway, a prominent stationary engine builder of the city. John Roach, perhaps the most famous iron shipbuilder of post-war Philadelphia, moved to that city after serving his apprenticeship in New York under James P. Allaire.[4] His exodus from New York was not surprising for while New York City continued to be the center of traditional wooden ship construction, the building of iron ships along the Delaware River was earning it a sobriquet as the "American Clyde."

LEADING PHILADELPHIA ENGINEMAKERS–1838[5]

Enginemakers	*Number of Engines Operating in Philadelphia in 1838*
Rush and Muhlenberg	26
Levi Morris	24
Hyde and Flint	15
M. H. Baldwin	12
Stacy Costell	10
Thomas Halloway	7
James Brooke	6
Daniel Large	5
Prosper Martin	4
James T. Sutton	4

Starting in the 1830s, Philadelphia enginemakers also gained a considerable reputation for the construction of locomotives. This success can be traced to the wide knowledge of stationary engines already possessed by local mechanics. Oliver Evans' early idea of building workable steam carriages for use on common roads had never completely died. About 1828 a group of promoters built a carriage at the machine shop of Nicholas and James Johnson in Kensington. This low-powered vehicle had a single cylinder, set horizontally, with connecting-rod attachment to a single crank at the middle of the driving axle. The machine was abandoned after a faulty steering device caused it to careen into a shop window.[6]

The real stimulation to activity, however, was the announcement in 1831 that trials had been scheduled to select a locomotive for Baltimore and Ohio Railroad service. A number of makers of stationary engines in Philadelphia entered the competition. Stacy Costell, already famous for engines he had built with vibrating cylinders, now built a locomotive with two six-inch cylinders of twelve-inch stroke. It was tried out later on the Columbia Railroad, after which it was broken up and the boilers set to work supplying a stationary engine. Thomas Halloway, another prominent builder of stationary engines, also began a locomotive for the trials, but it was never finished.[7]

The most famous Philadelphia engine builder was Matthias Baldwin. Starting as an apprentice with the jewelry firm of Woolworth Brothers, he later became a machinemaker specializing in engraved copper rolls for calico printing. Baldwin made an engine for his own use in 1827 and then built engines for sale; this trade "became extensive and profitable." He became interested in the Baltimore and Ohio trials of 1831 and later became one of the world's leading manufacturers of locomotives.

Joseph Harrison, another leading Philadelphia locomotive builder, also had a background in stationary steam engine construction. At the age of fifteen he was apprenticed to an engine builder named Frederick D. Sanno. When Sanno's business failed, Harrison went to work for James Flint of the firm of Hyde and Flint. Before his indenture had run out, he became foreman of this shop at the age of twenty-two. After working for a time with the locomotive builders and innovators William Norris and Colonel Stephen H. Long, he was hired as foreman by Garret and Eastwick, another Philadelphia engine-building firm. He became a partner in the enterprise, which soon began building locomotives. Between 1844 and 1851 he built railroad equipment in St. Petersburg, Russia. Samuel Wright, who had been an apprentice with Harrison in the shops of Flint and Hyde, joined with the stationary enginemaker James Brooke to make two experimental locomotives.[8]

Other Philadelphia experimenters built one or more locomotives during this same period, but the construction of railroad engines quickly became a separate branch of the machinemaking trade. Most of those stationary enginemakers who took an early fling at locomotive construction came to realize that there was more to it than merely placing an engine on a cart and soon went back to their familiar work. The industry, nevertheless, took firm root in Philadelphia, and Harrison wrote in later life that the city's workers "can be found everywhere, and for nearly forty years Philadelphia skill has been sought for to fill responsible places in all parts of the United States, in the West Indies, in South America and in Europe, and even in British India."[9]

To a lesser degree, stationary enginemakers in other cities went through the same pattern of trying their hands at building locomotives. In Boston, Holmes Hinkley made the transition smoothly and gained a wide reputation for the excellence of his engines. The pioneer Pittsburgh firm of McClurg, Wade & Company built five locomotives between 1834 and 1837, but then relinquished this line of work. Minus

Ward of Baltimore saw in the locomotive a possible use for his patented rotary engine and conducted several trials before giving up the enterprise. In New York, William James, a large-scale maker of stationary engines, experimented in 1833 with a steam carriage, "Calculated to run on a common turnpike road, also adapted to railroads."[10] They, like most of the Philadelphia makers, soon gave up building this type of engine.

Unlike Philadelphia, New York City did not become a center of locomotive building, but it continued to produce large numbers of marine and stationary engines. Charles Stoudinger's former partner James P. Allaire, who upon Fulton's death purchased his works, continued for many years to be one of the leading enginemakers of the city. His business increased so fast that during the 1820s he purchased some eight thousand acres of timberland at Howell, New Jersey, to ensure a supply of charcoal iron for his enterprise. After an initial success, this proved to be an ill-advised move when, in the 1840s, Pennsylvania iron made with anthracite gained wide preference for use in general machine work.[11]

Allaire's New York works employed approximately six hundred persons in 1831 and by 1847 his plant produced two hundred thousand dollars worth of iron goods. It continued in the manufacture of steam engines and boilers, heavy machinery, and a general machine business, but by 1853 it was principally engaged in building engines for ocean, lake, and river steamers.[12]

Robert McQueen's engine shop was still operating in New York in 1832, but its size was eclipsed by that of Paul A. Sabbaton. By 1810 the latter had been given "general superintendence of the late Mr. John Youle's foundry establishment" in New York City where, as he later remarked, he had witnessed the "continual exertions" of Robert Fulton to "obtain workmen who understood to work from drawings." From his early experience during the heroic period of steam engineering in America, Sabbaton went on to found his own extensive engine-building firm, and by 1838 his engines were operating in mills from New England to the Gulf States.[13]

Other prominent New York firms at midcentury included the Archimedes Works, started in 1833, and Hogg and Delamater's Iron Foundry, begun in 1835. The former, which operated at two sites in the city, employed 220 workmen and manufactured annually $235,000 worth of engines, sugar mills, dredging machines (of "improved" design), and "iron vessels of various descriptions." Hogg and Delamater's foundry employed 150 workers and produced each year $250,000 worth of engines and other machinery.[14] By 1853 these firms had been joined by the Morgan Iron Works and by Guyon, Boardman, and Company. Guyon had been a leading engineer with the Morgan firm before building his own works, which were housed at the foot of Eighth Street in a building 200 by nearly 100 feet and three stories high. This establishment, exclusively for the manufacture of steam engines,[15] contained a brass foundry, machine shop, blacksmith's shop, and storage space. By 1855 a total of 3,130 men were employed in seventeen engine-building firms in the city, and the largest of them was the Novelty Works.[16]

During the late 1820s Dr. Eliphalet Nott, president of Union College at Schenectady and an indefatigable inventor, had begun work on an ill-fated steamboat named *Novelty* because the arrangement of its machinery was a novelty. In this work he enjoyed the aid of Hezekiah Bliss—friend of Fulton, partner of Daniel French, and steamboat builder and captain on his own account.[17] The engine for the experimental vessel was built at a foundry on the East River in New York City. The foundry later was named the Novelty Works after Nott's boat and

The Novelty Iron Works, circa 1840, located at the foot of 12th Street and the East River, New York City. To the left are the boiler shop, smith shop, and machine shop. At the rear is the iron foundry, and at the right are the pattern house, the paint shop, and two unidentified buildings. Compare this view with the ground plan of the works a decade later.

was supervised by Bliss for some years. The works continued to operate until the wearing out of the machinery and the sharply rising land prices in the area led to abandonment of the site in 1870.

New York City did not contain all the engine shops in the state. Just as the cities lost their monopoly on the use of engines, so urban manufacturers along the Atlantic coast had to share their markets with builders in the interior. By 1855 there were twenty-eight engine builders scattered throughout the State of New York.[18] The manufacture of engines in Buffalo, for example, began when Samuel Wilkeson erected his Buffalo Steam Engine Works about 1825. As a youth he had

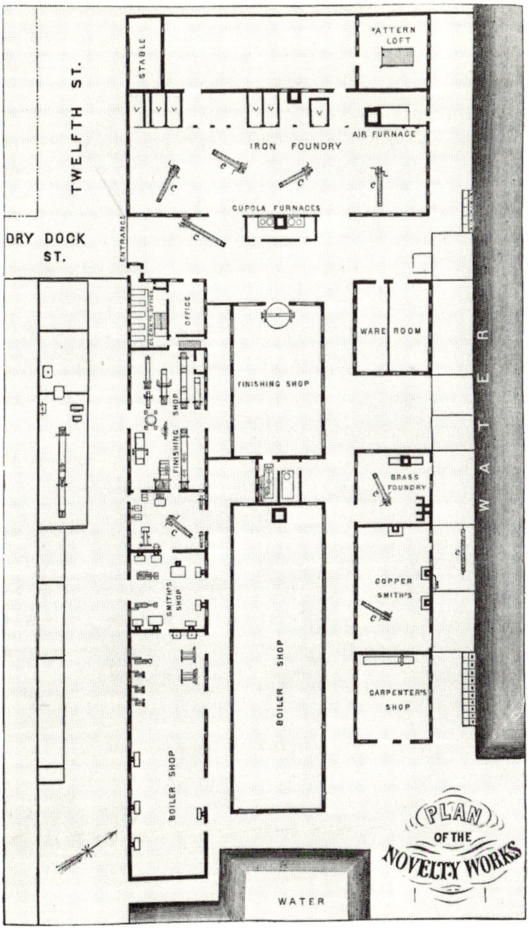

Ground plan of the Novelty Works, New York, 1851. From *Harper's Magazine*, volume 2 (May 1851), page 723.

Lathes in the machine shop of the Novelty Works, New York, 1851. From *Harper's Magazine*, volume 2 (May 1851), page 732.

lived near one of the first blast furnaces in Ohio, where he had learned the iron business.[19] By 1855 Buffalo contained a total of four steam engine and boiler works.[20]

One of the most extensive and comprehensive engine shops in the State of New York was the Matteawan Machine Works, located on the Hudson River below Poughkeepsie. This was one of the most complete machine shops in the country, and was set up to produce every part of a cotton mill from the water wheel or steam engine, to the card, spindle, and loom. These works did a large business in supplying machinery to southern manufacturers. A visitor wrote in 1850, "It is the custom of the Matteawan Co. to furnish drawings and plans of factories with all the necessary machinery. Those who are about starting factories receive all the information they want, which enables the stockholders to be posted up on all the business before him. All the working plans are given and thus, those who desire to enter upon the business of manufacturing need not be ignorant of the expense." At the time of this visit, the equipment being built included: six large steam engines in various stages of completion; an engine and two hundred looms for a cotton factory in Mobile, Alabama; an engine and four thousand spindles for a cotton factory in Greensboro, Georgia; and machinery for three water-powered factories in Tennessee. It was reported that during the previous year the company had built "63 engines, averaging 100 horse power each. These have been built for different parts of our country, and other countries. One splendid engine of 125 horse power is now on hand, for a large flouring mill in Havana, Cuba."[21] The Matteawan Company maintained an agent in New York City, through whom it sold machine tools and locomotives, as well as stationary engines and complete factories.

One of the most successful engine shops in New Jersey was the Speedwell Ironworks at Morristown. It was here that Stephen Vail, who later became a backer of Samuel F. B. Morse's electric telegraph, built the engine for the famous *Savannah* which, in 1819, was the first steam-powered ship to cross the Atlantic. The works, capitalized at $13,000, produced $15,000 worth of steam and other machinery annually. In 1820 Vail listed his machinery as "2 Trip hammers, 3 Turning Lathes, 1 Boaring Mill, 1 Engine for Cuting large Screws, 2 Blade Mills, 6 Pair Bellowses, 5 Blacksmiths Fires, 4 Water Wheels, &c." During the 1840s and 1850s New Jersey, too, saw an increase in the number and size of machine shops producing engines and other materials. These establishments included the Jersey City Iron Foundry, which employed fifty men in 1846; Vancleve and McKean, in Trenton, producers of steam engines, water wheels, mill gearing, and related items; and Watts and Belcher, in Newark, who in 1851 turned out "Steam Engines, Lathes, Planing Machines, Power Presses, and Mechanics' Tools of all descriptions."[22]

Engine building in the West continued to grow as the population and the manufactures of that section expanded. It had begun in Pittsburgh during the War of 1812 and soon spread to Cincinnati and Louisville. Cincinnati by the late 1830s had fourteen iron foundries and engine shops, as well as ten machine shops. It was destined to become one of the nation's great centers of machine tool trade. The Hamilton Foundry of Harkness, Voorhees and Company and the Franklin Foundry of Yeatman, Wilson, and Shield were among the most extensive engine works in the

West.[23] Louisville also developed into a major engine-building center. By 1839 the city's shops had produced over 244 engines, most of which went into western and southern mills. Beatty and Curry's Jefferson Foundry, Schnetz and Baldwin's Washington Foundry, and the Louisville Foundry of Bridgeford, Ricketts, and Company were among the foremost shops in the city.[24]

In 1837 Zanesville, Ohio, had two steam engine and machine shops, the most notable being Dare, Whitaker, and Company. Wheeling, Virginia, had at least three engine shops in the late 1830s, the most important of which were the establishments of Helm and Richardson and Arthur M. Phillips.[25] The sixty-one engines engaged in pumping salt water at the nearby Kanawah salines put a premium locally on steam engineering know-how.[26] The Mississippi Foundry of Gaty and Coonce built twenty-two engines in St. Louis, Missouri, in the year 1836, nineteen of which were stationary. This firm also did a large business in providing equipment for the various lead works at Galena. A second shop was established the following year when the St. Louis Foundry was begun by Bemis, Kingsland, and Lightner. Previously Bemis had had an engine shop in Pittsburgh.[27] By 1849 there was an engine shop in Indianapolis, and Detroit had three large engine works, besides two which specialized in making boilers.[28] Chicago had engine shops by 1852, and their products were being used in Wisconsin, Iowa, Michigan, and Indiana, as well as in Illinois.[29]

Pittsburgh remained the engine-building center of the West. Forty-four individuals or firms are listed as having made the 133 engines operative in the city in 1838.[30] The largest number of engines was credited to Litch, Sinamon, and

LEADING PITTSBURGH ENGINE AND MACHINE SHOPS–1857[31]

Name of Firm	*When Established*	*By Whom*
Knap, Wade, & Co.	1814	Joseph McClurg
A. Irwin & Co.	1820	Arthurs & Benny
John B. Warden & Son	1830	Columbia Steam Engine Co.
F. & W.M. Faber	1834	E. & F. Faber
Robinson, Minis & Miller	1836	
James Rees	1840	R. Wightman
W.P. Eichbaum	1840	
J. Tomlinson	1841	J. Tomlinson
H. Wightman	1844	H. Wightman
Haight, Hartupee & Co.	1847	White, Hartupee & Co.
Ramsey & Renton	1847	R. Ramsey & Co.
S.S. Fowler & Co.	1848	S.S. Fowler & Co.
C. Preston	1848	
Cridge, Wadsworth & Co.	1855	Cridge, Wadsworth & Co.
Graham, Bole & Co.	1856	Graham, Bole & Co.
Robert Lea		

Company, eighteen; F. A. Bemis and Company, eleven; McClurg, Wade, and Company, nine; John Arthurs, eight; Samuel Smith, eight; and William Johnston, five. Business partnerships were formed and dissolved so frequently that it is impossible to measure the influence of any one individual. The recurrence of certain names, however, makes it clear that there was a good deal of continuity in local engine building. For example, the firm of Knap, Wade & Company, operating in 1857, traced its pedigree, through Wade, back to Joseph McClurg. A. Irwin and Company, also operating in 1857, was using the foundry that had its origin in James Arthurs' early woolen mill and steam engine works.

The Pittsburgh engine shops at this time were typical of their kind as they also manufactured a general line of ironwork. In 1836 Bemis and Company built twenty-six engines (all but four of which were stationary) in addition to "6 planeing machines, 2 condensers, 1 carding machine, 1 jack of 120 spindles, 1 hemp picker, 1 wool ditto, 1 satinett loom, 2 framing machines, &c. all amounting in value to about $60,000." In that same year, McClurg, Wade and Company claimed to be "the only foundry to the west of the Allegheny, at which locomotive steam engines, heavy cannon and field artillery, are made." McClurg had made cannon for many more years than he had built engines, receiving his first ordnance contracts during the War of 1812.[32] British industrialists, visiting American ironworks in 1854, noted "the practice which prevails of combining various branches of manufacturing in the same establishment." They explained that "the policy of thus mixing together the various branches arises, in addition to other causes, from the fact that the demand is not always sufficient to occupy large works in a single manufacture."[33] This diversity of product and the roving nature of American mechanics probably account for the many innovations of these early shops.

The building of steam engines was slower in becoming established in New England, but by 1838 that section was already well supplied with builders of reputation. Enginemaking began in Boston in the 1820s: the observant visitor Anne Royall mentioned a steam engine manufactory in the city in 1826 but did not identify it by name.[34] It may have been the establishment of E. A. Lester who, after a long apprenticeship in other cities, built a number of the earliest engines in Boston. His "first experience in steam engines," he wrote, "commenced in 1814, in the city of Baltimore, where I was employed by Capt. George Styles in the manufacturing of the rotary steam engines" which the latter had patented. "Since leaving Baltimore," he continued, "I have been employed in New York, North Carolina, and this city [Boston], in the building and operating of both high and low pressure" engines.[35] By 1838 there were eight engine shops in the city, the largest of which were those of Hinkley and Drury, J. and Seth Adams, Thomas Ashcroft, James M. Pond, C. Alger and Company, and Otis Tufts. The Massachusetts census of 1855 showed that there were forty-four engine shops in the commonwealth, seventeen of which were in Boston.[36]

During this time the most important engine builder in Boston was Holmes Hinkley. So extensive was his trade that by 1838 he had made nearly a third of all the engines in Boston, as well as a fifth of those in the rest of the state, and half of the engines in Maine. Born in 1793, Hinkley was apprenticed to a carpenter when he was fourteen. He traveled in the South for some time, returning to Boston during the slack seasons. In 1823 he went to work in a machine shop, where he not only learned the trade, but soon became foreman of the shop. When the firm failed, he set up his own general machine shop. He built his first steam engine in Boston in 1826. Two years later, in addition to his usual line of machinery for ropewalks, Hinkley began to manufacture engines in quantity. As business increased, he purchased adjoining land and accepted two partners, the first of whom was Gardner P. Drury. Hinkley claimed that he "built most of the steam-engines then in use in New England."[37] Having built his first locomotive in 1840, he entered the business in earnest in 1847 and eventually acquired a considerable reputation in that line. Active control of the business passed from his hands in 1857, and shortly thereafter the firm went bankrupt. Before he died in 1866, Hinkley managed to recoup some of his fortune by taking government contracts for shot and shells during the Civil War.

In the late 1820s Daniel Copeland initiated the making of steam engines in Connecticut.[38] His first order was from a foundry in Hartford. His son, Charles W. Copeland, later remembered that "my father went to New York to get men to go to Hartford to work on it as there were no men in Hartford who had any experience in that class of work."[39] Charles Copeland himself later had a distinguished career as a steam engineer. After studying at Columbia College, he became superintendent of the West Point Foundry in New York, chief steam engineer of the United States Navy, superintendent of the Allaire Works, and finally an independent "Consulting and Mechanical Engineer, Surveyor of Steam Machinery, etc."[40]

Daniel Copeland's need to search for steam engine builders in New York City was indicative of the ignorance of that art in Connecticut. In 1838 Copeland had built only four of the engines then running in the state. Half of the others were built in New York State, principally by Paul A. Sabbaton and William T. James. The leading engine builders in Connecticut, aside from Copeland, were the Middletown firms of Asa Richardson, and Guild & Douglass. Connecticut enginemakers were subjected to technological imperialism from both north and south. John Gore built engines in Brattleborough, Vermont, and sold them along the Connecticut River Valley. His particular territory was in western Massachusetts, where about eighty miles of overland transport insulated him from Boston competition, but a few of his engines found their way into Connecticut.[41] Despite such incursions of out-of-state engines, Connecticut supported six engine shops in 1845-46.[42]

Rhode Island was the earliest seat of large-scale manufacturing activity in New England, and it was also the first state to adopt the use of steam engines. The Hope Furnace of 1780 had one of the earliest engines in the country, and after its construction the area was never without some form of steam power. Perhaps the first serious enginemaker was Robert L. Thurston, who became associated with the steamboat inventor John Babcock, Sr., in 1825. At Providence the elder Babcock was constructing a boat, the most controversial feature of which was the intended replacement of the normal boiler with a so-called "steam generator,"[43] which was to heat small amounts of water frequently rather than a large amount constantly.

Sometime after the successful completion of the Babcock boat in the fall of 1825, Robert Thurston moved to Fall River, Massachusetts, where he engaged in the iron business.[44] Returning to Rhode Island in 1830, he joined with John Babcock, Jr., in establishing what later became the Providence Steam-Engine Works.[45] The firm prospered, although it went through the usual series of evolutions in partnership. By 1838 it had captured the largest share of the engine business of the state, providing half the engines then at work in Newport, and seventeen of the twenty-three engines in Providence.

The firm of Thurston, Greene, and Company, after some initial difficulty, gained the reputation of building "a standard form of expansive steam engine."[46] Samuel Slater used a Babcock and Thurston engine in his cotton mill in 1835 and complained to its makers that "the valves of our Engine are not lifted equally one sett being raised at least a quarter Inch more than the other." Slater pointed out that besides the waste of steam caused by this defect, the engine lacked that "equal motion [which] is nearly indispensable to our operation." He concluded, "(with regret that it Cannot Consistently be avoided) that the rods do not Satisfy us as they are at present."[47] Perhaps prompted by this and similar complaints, the Providence Steam-Engine Company purchased the right to produce the "cut-off" valve patented by Frederick Sickles in 1841. This adoption led the Providence firm into a long series of legal entanglements with George Corliss, whom they charged with infringement.

Robert L. Thurston died in 1874, eleven years after he had been forced to retire from the engine business. His firm–no doubt in company with many other major machinemaking establishments–was caught at the outbreak of the Civil War with a large amount of credit extended to the southern market.[48] As one eulogist commented, "Mr. Thurston was not fortunate in monetary matters."[49] His son, Robert Henry Thurston, later became an outstanding professor of engineering at Stevens Institute, in Hoboken, New Jersey, and at Cornell University. He was also author of an important history of steam engines and translator of Sadi Carnot's *Reflections on the Motive Power of Fire.*

Providence, Rhode Island, produced a remarkable group of steam engineers in addition to the Thurstons; the best known was undoubtedly George Corliss. The presence of aggressive manufacturing activity and technologically sophisticated machinists in Rhode Island no doubt created a context conducive to innovation generally, while the early and continued dependence upon steam was able to give direction to improvement. The number of engines in Providence rose steadily, from twenty-three in 1838, to sixty-five in 1850, and to eighty by 1856. *Scientific American* acknowledged this progress at midcentury with a: "Well done little Rhody."[50]

The southern market for steam engines remained tributary to northern and western shops until well into the nineteenth century. Among southern cities, as late as 1838, only Richmond, Virginia, along the southeastern seaboard, and New Orleans on the Gulf, had active steam-engine shops. The Eagle Foundry, established at Charleston, South Carolina, about 1837, concentrated on engine building soon after it began operations. This city had one foundry in 1802 and eight in 1858. By

1860, only New Orleans, Richmond, and Nashville, in the South, surpassed Charleston in total foundry production.[51] In 1842, Niles' *Weekly Register* spoke glowingly of "some splendid engines" having been made recently in South Carolina, but it was not until the decade before the outbreak of the Civil War that southern shops began to manufacture engines in any quantity or with any regularity.[52] In 1850 engines were being built in Columbus, Georgia, and the following year the Montgomery Manufacturing Company, of Montgomery, Alabama, advertised that it was willing to furnish steam engines, sugar mills, and all sorts of iron and brass castings, "upon terms as favorable as can be secured from the best northern establishments."[53] In 1852 a foundry in Mobile, Alabama, was producing engines "which, for strength of construction, finish, and perfection of movement, are equal to any work of the kind ever made in Northern or Western foundries." This firm was so far developed that it was even exporting machinery to Mexico.[54] By this date it was apparent that growing prosperity and nationalism were beginning to overcome the handicaps which initially had retarded the building of steam engines in the South. It seems probable that, for the South as a whole, the manufacture of steam engines was retarded more by inadequate financial resources than by a hereditary technological ineptness.

The export of machinery—especially of steam engines—marked a new maturity for American machinists. Dependent at first upon the technological knowledge of English steam engineers, and even faced with a certain amount of direct competition from English engines, American builders were in a position to export their own handiwork by the 1830s. As early as 1827 it was reported that a Baltimore manufacturer had constructed and sent to Holland a "steam flouring mill, on the American plan."[55] While this was perhaps an early and unusual example, by 1832 both Mahlon Rogers and McClurg, Pratt, and Wade of Pittsburgh exported steam engines on occasion: the former to Mexico and the latter to South America.[56] New York's Matteawan Works in 1850 exported a "splendid engine of 125 horse power . . . for a large flouring mill in Havana, Cuba."[57]

Engineers as well as engines were sent abroad. About one hundred engineers commuted between Boston and Cuba in 1851. The claim was made that "there is in nearly every plantation in Cuba, a sugar mill driven by steam engines, built usually in New York or Boston." Cubans were thought "not qualified for the situation" of engineer at these mills, so every October, Boston mechanics sailed for the island, earned from one hundred to one hundred and fifty dollars a month during the sugar season, and returned to the more salubrious climate of Boston in June.[58]

Such birds-of-passage were not unusual. Upon the death of one steam engineer, who was born in 1837, it was remarked that "he had the experience of every one who at that early day desired to enter the ranks of the engineers, taking his earlier training in the form of experience in the management of steam engines in out-of-the-way places where no one else could be found to do the work."[59] For the less adventuresome, an equally respectable way to learn the trade of engineer was to spend some time as a workman in an engine shop.

A young man identified only as "W. F., of Ohio," wrote to the editor of the *Scientific American* in 1853, asking how he could become a steam engineer. The editor answered: "If you have any taste for the engine you could soon learn to operate one: a few months practice in an engine building establishment would be better for you than all the books you could purchase." He added that "careful and experienced men are wanted as engineers; your age is no objection."[60] Men could sometimes move in the opposite direction. Erasmus W. Benton, proprietor of the New Albany (Indiana) "steam engine and sugar mill establishment," in 1830, had gotten his start on board the steamship *Enterprise* in 1815.[61] There is evidence, however, that a certain gulf remained between the building and operating of engines. One wandering machinist, who had been in the employ of Robert McQueen, James P. Allaire, Paul A. Sabatton, and the West Point Foundry, all in New York, was engineer on board the ill-fated *New England* the night the ship blew up on the Connecticut River. He was unable to prevent the accident that resulted in a loss of fifteen persons killed and ten injured. Indeed, a correspondent informed a special committee of the Franklin Institute in 1830 that "the greater part of the engineers and officers on board of steamboats whom I have met with, I have found quite ignorant of the nature of the power they are using. The former generally acquire their responsible profession at the engine factories, or in the duties of firemen. It is not supposed, therefore, that they should be acquainted with the *science* of steam."[62] Whether or not machinists made good operating engineers, men continued to move from the factories to the boats.

The typically informal and uncontrolled education of steam engineers seemed irrational and repugnant to some observers. As within any maturing trade, there were calls for professionalization and standardization of training. One of the most comprehensive, though generally ignored, suggestions was that the federal government establish an academy of steam technology. Cadets would divide their time between classes in science and practical instruction in a complete engine shop. Upon graduation, they would be certified as engineers, well trained in "the art of manufacturing, and the manner of working steam-engines." Engines were to be made by the cadets, and "to avoid the censure or injustice of the nation founding a monopoly in the manufacture of steam-engines, the price of all such as are not expressly made for the use of the government shall be regulated by the prices at which like engines are sold by the persons already, or shall hereafter be, engaged in manufacturing steam-engines; and no engine shall be repaired at the national factories, except such as are the property of the government." Finally, it was suggested that as the contemplated school "is one of general benefit to science and the country, it is believed to be a proper object for the appropriation of the Smithsonian legacy."[63] Needless to say, this proposal was rejected, and engineers continued for many years to be trained willy-nilly.

Perhaps the most important difference between the early and later engine shops was in the amount of individual skill and resourcefulness which went into the making of each engine. An engine built by Nicholas Roosevelt in 1800 would have been largely handmade; that is, each part would have been separately and newly designed and fitted to the rest. By the middle of the century, this was no longer the case. Any given piece of an engine would be, for practical purposes, identical to (though not necessarily interchangeable with) the corresponding piece of any other engine of the same power made by the same firm. As early as 1819 it was reported that in England, "the great demand for these machines, which has taken place since their value has been so fully understood, has occasioned them to be manufactured in the large way by several engineers, who adopt the same system as is pursued in making watches and clocks, *viz.* that of having the workmen instructed in making the separate parts, and employing machines and tools for every operation which admits of such aid."[64]

The uniformity of corresponding parts was inherent in the building of steam engines from the beginning of their production for the market. Many engine parts were cast from patterns, and it was simply sensible business practice to use a pattern already made whenever possible. An increased volume of production brought this process very close to the system of interchangeable parts. Benjamin Henry Latrobe saw clearly the advantage of building many engines of the same size. He estimated that a one-foot cylinder engine could be built in 1812 for $2500. There would be no profit at this low price, but if another was made from the same patterns, the five to six hundred dollars spent on patterns for the first engine could be saved, thereby becoming profit.[65]

Another machine builder had the same idea, but expressed it differently. Reminiscing about Philadelphia machine shops of the early nineteenth century, Coleman Sellers told of an old time engineer who "used to say that the jobber, selling his thoughts to one purchaser at a time, wastes his brain capital. He spends his energies on each emergency with little or no prospect of repeating the construction upon which he has expended much thought or of applying again ingenious devices worked out for unique cases. The manufacturer on the other hand, invests his mental capital, in what, if salable, yields him a constant income."[66] Sellers' description of this old mechanic idealized the best of the type. "He was," wrote Sellers, "a walking encyclopedia of his own and others' experience. He could remember all sorts of devises contrived by this or that man, and was full of anecdotes to illustrate his points. His drawing board was any plank at his command and his instruments a bit of chalk. He knew where every pattern was placed in the loft and could combine them to meet his emergencies."

The early steam engines were custom made. The most important of several reasons for this were: the engine was built to fit the particular job for which it was ordered; the purchaser, through the intercession of an independent engineer like Latrobe, or on his own initiative, would prescribe specifications; and, frequently, an idea for an improvement by the maker would dictate some change in the normal design. All of these militated against a uniform product, but time would almost necessarily remove such divisive elements; it would bring an increase in the number of engines ordered, as well as a kind of technological inertia.

It was noted in 1859 that among the dozen or so engine shops in Philadelphia, "there are none in which the Steam Engine is an exclusive article of manufacture, or none which keeps a large stock of finished Engines constantly on hand. The necessity for anticipating orders . . . has not hitherto been felt by the makers, their facilities being such as enable them to meet the demand, as it arises, with sufficient expedition."[67] This, however, was not universally the case. William Burdon, one of the leading engine builders in Brooklyn, New York, advertised in 1848 that he had "in process of manufacture 150 Steam Engines from 5 to 20 horse power." That this was not a unique occasion is evidenced by his advertisement eight years later that he had "50 STEAM ENGINES–From 3 to 40-horse power."[68] Aaron Kilborn, a Connecticut engine builder, advertised in 1851 that he was "finishing four 14 horse engines" as well as "several 6 horse engines" and "several of smaller capacity, complete."[69]

Whether complete engines were kept in stock was not so important as whether patterns were kept and used again. This practice was followed even in Philadelphia, where it was said of the firm of Reaney and Neafie that "having made it a rule to preserve all patterns, their stock at present is very large."[70] The preservation of patterns, when coupled with the logic of an ever-increasing volume of engines constructed, would lead to the manufacture of a uniform product, with more or less interchangeable parts. It took no flash of genius or bold innovation to bring this about in the steam engine trade.

Indeed the impact of steam engine construction on the whole evolution of American machinemaking can hardly be overemphasized. An anonymous reviewer writing in 1822 noted that "the improvements in manufacturing the metals, particularly iron, have been very great, since the date of the steam engine." At the beginning of the century, he recalled, "our iron furnaces produced no castings, which united the properties of strength and solidity, with such softness as to enable the workmen to give them any kind of finish." This backwardness in casting had been, in his opinion, the prime cause of America's lagging behind England in the art of machinemaking and the major reason for our machinemakers so often turning to wood as a material.[71]

In the United States not only was the steam engine the first widely manufactured machine which of necessity had to be made of iron rather than wood (early tcxtile machines were more often made of wood and brass), but it also continued to be the most important. The 1860 census showed that the machine-building industry in the United States accounted for only about three percent of the total value of manufactured goods, and steam engines accounted for 90 percent of all machinery production. The influence of steam engines was obvious and decisive.

During the first half of the nineteenth century, the making of steam engines progressed from adventure to maturity. By the 1850s engine shops of various sizes, while still concentrated in the larger cities of the nation, were to be found in all sections of the country. The knowledge of steam technology had become so widespread that nearly any town could boast of some mechanic able to build an engine, and at the same time, the leading engine shops became ever larger, though not more specialized. Soon the time came when American engine builders found it possible to export their handiwork to South and Central America, and even American engineers found a call for their talents abroad. Finally, the increasing demand for engines made it mandatory that standardization replace improvisation in their construction. By midcentury, the building of engines was probably no more hazardous or speculative than any other type of manufacture in America.

Notes

[1] *Scientific American,* volume 10 (16 June 1855), page 315.

[2] John Fritz, *Autobiography of John Fritz* (New York, 1912), pages 32-33.

[3] *A. M'Elroy's Philadelphia Directory, for 1840* (Philadelphia, 1840), passim.

[4] "John Roach," *Appletons' Cyclopaedia of American Biography,* volume 5, page 268.

[5] Woodbury Report, page 156.

[6] Joseph Harrison, Jr., *The Locomotive Engine, and Philadelphia's Share in Its Early Improvement* (Revised edition, Philadelphia, 1872), pages 19-20.

[7] Ibid., pages 30-31.

[8] See sketch of Harrison in *Dictionary of American Biography,* volume 8, pages 345-346; Harrison, *Locomotive Engine,* pages 56, 57.

[9] Harrison, *Locomotive Engine,* page 60.

[10] John H. White, *Cincinnati Locomotive Builders, 1845-1868,* U.S. National Museum Bulletin 245 (Washington, 1965), page 8; Minus Ward, *Remarks, Propositions and Calculations, Relative to a Rail-Road and Locomotive Engines to be Used upon the Same, from Baltimore to the Ohio River* (Baltimore, April 1827), pp. v-vi; Thomas Richard Thomson, *Check List of Publications on American Railroads before 1841* (New York, 1942), page 69.

[11] Sketch of Allaire in *Dictionary of American Biography,* volume 1, page 182; Brown, *Allaire's Lost Empire,* page 33; Morrison, *Steam Navigation,* page 180.

[12] *Niles' Weekly Register,* volume 40 (27 August 1831), page 452; *Hunt's Merchants' Magazine,* volume 16 (January 1847), page 93; *Scientific American,* volume 9 (17 December 1853), page 110.

[13] McLane Report, volume 2, page 115; Woodbury Report, passim; Affidavit of Paul A. Sabbaton, 19 December 1838, in *Heirs of Robert Fulton,* 27 January 1846, House Document 145, 29th Congress, 1 session, page 17.

[14] *Hunt's Merchants' Magazine,* volume 16 (January 1847), page 93

[15] *Scientific American,* volume 9 (17 December 1853), page 110; ibid., volume 8 (22 January 1853), page 149.

[16] *Hunt's Merchants' Magazine,* volume 37 (September 1857), page 382.

[17] Preble, *Chronological History,* pages 58-59. Bliss' account of this story is given in "Resolution of Mr. Beekman on the Petition of Levinus Vanderheyden, Neziah Bliss and Joseph D. Monnell," *Documents of the Senate of the State of New-York, Seventy-Sixth Session. 1853,* Document 68, volume 2 (Albany, 1853), pages 3-5.

[18] This figure was a result of a state census in that year. It does not make clear whether it included the city of New York. *Hunt's Merchants' Magazine,* volume 39, (September, 1858), page 382.

[19] John C. Lord, "Samuel Wilkeson," *Publications of the Buffalo Historical Society,* volume 4 (1896), page 76; Samuel Wilkeson, Jr., "Biographical Sketch," ibid., volume 5 (1902), page 143.

[20] *Hunt's Merchants' Magazine,* volume 38 (June 1858), page 691. See also [Joseph] Whitworth and [George] Wallis, *The Industry of the United States in Machinery, Manu-*

factures, and Useful and Ornamental Arts (London, 1854), pages 4-5.

[21] *Scientific American,* volume 5 (27 April 1850), page 253.

[22] "Stephen Vail," *Appletons' Cyclopaedia of American Biography,* volume 6, page 220; David Budlong Tyler, *Steam Conquers the Atlantic* (New York, 1939), pages 7-8; Return of Stephen Vails, no. 70 (New Jersey), schedules of the 1820 census of manufactures, record group 29, National Archives; *Hunt's Merchants' Magazine,* volume 15, (December 1846), page 598 and volume 18 (April 1848), page 444; *Scientific American,* volume 6 (19 July 1851), page 351.

[23] Lyford, *Western Address Directory . . . 1837,* pages 282-283; Charles Cist, *Cincinnati in 1841: Its Early Annals and Future Prospects* (Cincinnati, 1841), page 54.

[24] Woodbury Report, page 321; Lyford, *Western Address Directory . . . 1837,* pages 361-362.

[25] Lyford, *Western Address Directory . . . 1837,* pages 203, 218, 167-168.

[26] Woodbury Report, pages 226-229.

[27] Lyford, *Western Address Directory . . . 1837,* pages 403, 98.

[28] [E. Chamberlain], *The Indiana Gazetteer, or Topographical Dictionary of the State of Indiana* (Indianapolis, 1849), page 262; *Hunt's Merchants' Magazine,* volume 20 (March 1849), page 284; for Detroit see ibid., volume 35 (November 1856), page 556.

[29] *Hunt's Merchants' Magazine,* volume 26 (April 1852), page 509.

[30] Woodbury Report, page 191.

[31] Thurston, *Pittsburgh As It Is,* page 119.

[32] Lyford, *Western Address Directory . . . 1837,* pages 96-98; McLane Report, volume 2, pages 252-253.

[33] Whitworth and Wallis, *Industry of the U.S.,* page 3.

[34] [Anne Royall], *Sketches of History, Life, and Manners, in the United States* (New-Haven, 1826), page 342.

[35] *Reports and Documents Upon the Subject of the Explosions of Steamboat Boilers; Presented to the House of Representatives, 18 May 1832,* House Document 478, 22d Congress, 2d session (Washington, 1833), page 62. For Styles see *Niles' Weekly Register,* volume 17 (16 October 1819), page 114. For an illustration and description of Lester's pendulum engine see *Journal of the Franklin Institute,* volume 12 (August 1831), pages 98-104.

[36] Francis DeWitt, compiler, *Statistical Information Relating to Certain Branches of Industry in Massachusetts, For the Year ending June 1, 1855* (Boston, 1856), page 581; *Hunt's Merchants' Magazine,* volume 39 (July 1858), page 45.

[37] Quoted in Hinchman, *Hinkley,* page 23.

[38] His son thought this was in either 1826 or 1828. Hutton, "First Engines," *Transactions of the American Society of Mechanical Engineers,* volume 15 (1894), page 985.

[39] Quoted in ibid., page 985.

[40] "Charles Copeland," *Dictionary of American Biography,* volume 4, page 423; *Scientific American,* volume 7 (3 January 1852), page 127.

[41] See, for example, Woodbury Report, pages 47, 82.

[42] *Hunt's Merchants' Magazine,* volume 15 (December 1846), page 568.

[43] See Joseph H. Patten, "Notice of the Steam-Boat Babcock," *American Journal of Science,* volume 12, no. 1 (1827), pages 115-118.

[44]For the *Babcock* see *Niles' Weekly Register,* volume 29 (24 September 1825), page 56; ibid., volume 29 (8 October 1825), page 87; *American Merchants' Magazine,* volume 2 (15 October 1825), pages 173-174.

[45]Thurston, *Growth of the Steam-Engine,* pages 280-281.

[46]See obituary of Robert L. Thurston in *Scientific American,* volume 30, new series, (7 March 1874), page 145.

[47]S. Slater & Sons to John Babcock, 10 July 1835, letter book, volume 14, Slater Steam Manufacturing Company, Slater collection, Baker Library.

[48]William Frederick Durand, *Robert Henry Thurston; A Biography. The Record of a Life of Achievement as Engineer, Educator, and Author* (New York, 1929), page 33.

[49]*Scientific American,* volume 30, new series (7 March 1874), page 145.

[50]Woodbury Report, pages 88-89; *Hunt's Merchants' Magazine,* volume 34 (June 1856), page 675; *Scientific American,* volume 5 (20 July 1850), page 346.

[51]Ernest M. Lander, Jr., "Charleston: Manufacturing Center of the Old South," *Journal of Southern History,* volume 26 (August 1960), pages 334-335, 337.

[52]Volume 62 (20 August 1842), page 387.

[53]*De Bow's Southern and Western Review,* volume 9 (October 1850), page 431; *Scientific American,* volume 6 (19 July 1851), page 351.

[54]*Scientific American,* volume 7 (7 August 1852), page 376.

[55]Bishop, *American Manufactures,* volume 2, page 320. The "American plan" no doubt referred to the automatic features invented in the 1780s by Oliver Evans.

[56]McLane Report, volume 2, pages 240, 252-253.

[57]*Scientific American,* volume 5 (27 April 1850), page 253.

[58]Ibid., volume 7 (8 November 1851), page 59.

[59]Obituary of Joseph S. Ludlam, in *Transactions of the American Society of Mechanical Engineers,* volume 18 (1896-1897), page 1091.

[60]*Scientific American,* volume 9 (17 December 1853), page 111.

[61]Letter from Erasmus W. Benton, 7 August 1830, in *Communications Received by the Committee of the Franklin Institute, on the Explosions of Steam Boilers* . . . (Philadelphia, 1832), page 20.

[62]Woodbury Report, pages 458-459; *Communications Received* . . . , page 49.

[63]Woodbury Report, pages 250-252.

[64]"Steam-Engine," *The Cyclopaedia; or, Universal Dictionary of Arts, Sciences, and Literature,* volume 34 (London, 1819). For a description of the shops of Fenton, Murray, and Wood, of Leeds, England, see ibid., volume 34. Since many early engineers, both in America and England, were drawn from the ranks of clock- and watchmakers, this development is really not surprising. See A. E. Musson and E. Robinson, "The Origins of Engineering in Lancashire," *Journal of Economic History,* volume 20 (June 1960), page 222.

[65]B. H Latrobe to F. C. Graf, 5 May 1812, B. H. Latrobe MS. Coll., MHS. This opportunity was a major consideration in Latrobe's decision to go into engine building.

[66]Coleman Sellers, "President's Address, 1886," *Transactions of the American Society of Mechanical Engineers,* volume 8 (1887), pages 677-78.

[67]Freedley, *Philadelphia and its Manufactures,* page 316.

[68]*Scientific American,* volume 3 (8 April 1848), page 231; ibid., volume 11 (2 August 1856), page 375.

[69]Ibid., volume 6 (25 January 1851), page 151.

[70]Freedley, *Philadelphia and its Manufactures,* page 318n.

[71]"Improvements in Machinery," *North American Review,* volume 14 (April 1822), pages 411-412.

Chapter 8

Developing Engine Design

We cannot have discoveries every few days,
like those of the steam engine, the steamboat,
the spinning jenny. . . . The progress
of discovery is a gradual one: the
trimming off a superfluous shaft here, and
a wheel, crank or drum there, produces
important though not very striking results,
and upon such improvements in the
aggregate (and sometimes a very simple
one in the minutia of complex machines,)
depend the whole economics of the machinery–its
profits and losses.

SCIENTIFIC AMERICAN (1849)[1]

Even while they sought to master the techniques and to provide for the needs of engine manufacture, American steam engineers attempted to improve the product of their shops. Although the typical stationary steam engine in this country at midcentury was a considerably better product than its predecessor of fifty years before, it was not radically different from, or better than, an English engine of the same period. Improvement had gone on independently in the Old World and the New, but there was constant communication between the two: engineers and publications crossed the Atlantic in both directions and in increasing numbers. Added to this was the fact that the inherent logic of the steam engine tended to limit the potential number of improvements. The achievement of American engineers lay not so much in surpassing their English counterparts–although in some cases they did–but in succeeding in exploring and mastering an inherited technology which was itself in a process of evolution.

As soon as the Watt improvements were available in America, the old-type Newcomen engine was immediately abandoned. When Oliver Evans patented and introduced his high-pressure Columbian engine in 1804, the battle was joined between his partisans and those of Watt's low-pressure, condensing system. James Watt had considered the use of pressures above that of the atmosphere, but abandoned the plan.

While still an apprentice in the 1770s, Evans had witnessed the expansive power of steam but "formed no idea of the means of its application, until he met with a description of an atmospheric steam engine." He claimed to be "astonished to find that the steam was not applied as the prime mover, but only as the means to form a vacuum, to apply the weight of the atmosphere. This he conceived to be a great error."[2] Drawing upon his rather inaccurate knowledge of the history of the steam engine, Evans declared that "this invention and discovery of Newcomen's, made from the hints given by doctor Hooke . . . was the first step . . . from the simple path of nature! Here they lost sight of the true principle discovered by the Marquis of Worcester."[3] With a touch of bitterness, Evans confessed that "to philosophy we are indebted for many of our most useful discoveries; yet," he continued, "this single case should put our philosophy to the blush, and teach us, however learned we may be, to listen with the closest attention, even to what the most illiterate mechanician, who has taken the simple works of nature for his guide, may say."[4]

Evans stopped short of calling himself a babe and suckling, but the implication was clear: nature's plan and true proportion had been betrayed into the toils of sophisticated, but specious, reasoning. Evans, the simple American mechanic, had rediscovered the true path of advancement. Benjamin Henry Latrobe, who was anything but a simple mechanic, later claimed to have been asked by Evans why engines in England were not built to use high steam. Latrobe replied by citing the abortive efforts of Hornblower, Symington, and Murphy in this direction. He pointed out that these engines "had been found to have so many inconveniences attached to them" that each had been abandoned.[5] Evans, however, sought explanations, not precedents. Inveighing against "the mighty and *infallible* opinions of Englishmen," he condemned their belief that "no inventions, improvements, or original ideas, at least possessed of merit, can spring from any other quarter of the world, than the 'fast anchored isle'."[6] Evans' feeling that science had presided over the birth of the low-pressure engine but not that of his own has been perpetuated by later writers.

When Evans introduced his Columbian engine in 1804, it faced immediate and bitter opposition. Latrobe led the opposition and never became reconciled to either Evans or his engine. He based his position on the facts that the high-pressure engine was too sentitive to the pressure of steam in the boiler, thus moving "unequally & precariously," and the danger of working with "the enormous force of the Steam on the boiler." In contrast, he maintained, the Watt engine worked with "no inconvenience or inequality" and was "as safe as a clock."[7] Ironically, Latrobe finally recommended the use of high-pressure steam, but in engines not made on the Evans plan.

Evans always complained that he was forced to watch his improvements exploited, even while their validity was being denied. Latrobe was only the most obvious offender in this respect. Daniel French, the early builder of western steamboats, expressed surprise in 1814 that Evans should pretend to any improvement in his use of high steam. "For my part," he wrote, "I never save now heard of any Engine but what used Steam more or less above the pressure of the atmosphere."[8] In fact, French had already built at least one steamboat, equipped with a "Seesaw" oscillating engine, "worked with high Steam without a Condenser, with a double Circular boiler."[9]

Part of this opposition to Evans and his claims was based upon a real reluctance to employ high-pressure steam because of the great danger involved when raising steam in the crudely made boilers of that time. Some of the resistance, however, was brought on by Evans' claim to have patented the very use of high pressure, regardless of whether in an engine of his design or not. When Thomas Copeland began to build engines in Pittsburgh, "upon the principles of the celebrated WATT & BOLTON," in 1813, he maintained "that he cannot be prevented from using steam of any elastic force. It cannot be pretended," he concluded, "that this is a new discovery."[10] Evans, however, thought differently. He made a clear distinction between his patented steam engines and what he considered his patented principles, but continued to warn against the unauthorized use of either.

Evans' steam-engine patent, issued first in 1804, was extended by an act of Congress in 1815 and did not expire until 1825. By the latter date, however, high-pressure steam was widely used in steamboats on the western rivers and was almost universal among stationary engines all over the country. Appearing before the Select Committee of the House of Commons, on Steam Navigation, the American Consul at Liverpool testified that in the United States, "all condensing engines are called low-pressure." Even these, he claimed, were sometimes worked at a pressure of twenty pounds to the square inch.[11] "An American low pressure engine (so called)," it was said, "is very commonly of low pressure only by courtesy." In 1834 an observer in Cincinnati, Ohio, counting sixty engines operating in that town, noted that they were "*all* high pressure," and spoke for a whole section when he added: "low pressure engines are not in good repute in the west." By 1838 only a small number of low-pressure engines were listed as being at work in American mills.[12]

The second major development in American steam engines during the first half of the nineteenth century was acceptance of the principle that, for greater efficiency, steam should be used expansively. Once again, Watt had already anticipated this development, having used it as early as 1782, but it is not clear whether the first engines of his design in America adopted this practice. Oliver Evans recommended that his Columbian engines be driven by expansive steam, and in his handbook for engineers he included a table "showing the proper time to shut off the steam" on each stroke.[13] Latrobe, always ready to belittle Evans' work, commented that "Evans shuts it off, to save his Engines from destruction, (as do all the high steam Engine makers in England) and uses the expansive power alone & separately."[14]

It was Latrobe's opinion that Francis B. Ogden, nephew of the famous Governor Aaron Ogden of New Jersey, was the true inventor of the expansive use of steam.[15] In 1811 Ogden, with his uncle and Daniel Dod, had built a steamboat to operate between New Jersey and New York. This same year he drew plans for an engine of two cylinders working at right angles and using steam expansively; it was patented in 1813. According to the specifications for this patent, Ogden planned to use "two or more cylinders in such a manner as to form one engine, with the view to cutting off the steam, of whatever pressure, at 1/4, 1/3, or 1/2 the stroke of the piston, and using it expansively for the residue."[16]

Whether Ogden's engines were directly responsible for the acceptance of expansive steam in this country is perhaps a moot point. Latrobe built an engine in 1815, on Ogden's plan, for a woolen mill in Steubenville, Ohio. He called this engine "the most powerful Machine I have ever seen," and suggested to Ogden that it be worked "with high steam without an air pump."[17] The following year, Ogden placed one of his engines in a steamboat operating on the James River in Virginia, and in 1816 he made a journey to England where he met the venerable Watt. When asked to comment on the novelty of Ogden's engine, Watt answered that although the arrangement of the cylinders was new, "the expansive power of steam has been long used in Cornwall, where much larger engines are erected than are at first required: of course, they are not put to their full work; 1/4 or 1/3 of the power being sufficient, the cylinders are only thus far filled with steam, and the residue of the stroke is made by expansion."[18]

It has been reported that Ogden's patent was freely infringed on the western rivers, and that Murray and Company, of Leeds, England, after making an engine under Ogden's direction, used the patterns to produce others for the American market.[19] It may be assumed, therefore, that Ogden's principles were fairly widely known by the 1820s. Even so, it seems likely that, since the high-pressure engine became the common type for stationary purposes, it was more through Evans than Ogden that the use of expansive steam became widespread. At any rate, it was said in 1854 that "almost every engine is now so worked."[20]

A third area of improvement in American steam engines was the positioning of the cylinders. At the beginning of the century all engines used vertical cylinders with beams. By midcentury the cylinder was usually placed horizontally. John Fitch was the first American inventor to attempt to build an engine with a horizontal cylinder, planning to place one in his steamboat of 1786. When he informed Josiah Hornblower of this intention, the latter warned against it. Fitch had conceived that plan "for the following reasons, to prevent its being too top heavy, and that the jarrs that the Vessel would receive from the seas would not be so likely to effect the Works."[21] Hornblower responded with the warning that "the horizontal position of the cylinder may be proper for the reasons you mention, and also for the more convenient application of the power, yet I fear it will be apt to make a leaky Piston."[22] Whether convinced by these arguments or others, Fitch had abandoned the use of a horizontal cylinder by the time he built his second boat in 1788.

The horizontal placement of a steam cylinder permitted a direct linkage between the piston rod and an eccentric arm connected to the flywheel. This seemed the best method of avoiding the use of the bulky and expensive beam that was common to the Watt and the Newcomen engines. Within the limited space of an industrial establishment, such an arrangement was clearly to be desired, but for many years it was retarded by a fear that the weight of the piston would cause excessive wear along the bottom length of the cylinder. The horizontal cylinder was little used in England before 1825 and perhaps became more common in the United States at about the same time. By 1838 the distribution of engines in Philadelphia County

A typical horizontal steam engine, 1848.
From *Scientific American,* volume 4 (4 November 1848), page 54.

showed 2 rotary engines, 13 oscillating engines, 57 with upright cylinders, and 107 using horizontal cylinders.[23]

By 1852 it was said that "the favorite construction for land engines throughout the United States is that in which the cylinder is placed horizontally on flat beds."[24] It cannot be demonstrated that any one engine builder or inventor was responsible for this universal acceptance. Fitch's attempt to use such an arrangement was in no sense a precedent for later acceptance, but did show early recognition of its desirability. It is likely that with improved piston-packing and generally improved durability and accuracy of construction, builders felt more free to experiment with horizontal engines.

The oscillating engine gained considerable currency in America toward the middle of the nineteenth century. It too had originated in England, where William Murdock had made a model of such an engine while in the employ of Boulton and Watt. It was revived in 1827 by Henry Maudslay, but was used little for stationary engines in Great Britain. The first American use of this type of engine was made by Daniel French on his steamboats. Commenting on one of French's boats in 1812, Latrobe noted that "the horizontal Cylinder has been very often used, tho' not made to vibrate." He went on to praise French as "a man of very uncommon talents, both as to the conception & execution of anything to which mechanical power can be applied," but insisted that "the immediate application of the Piston to the Crank of the wheel which is rendered practicable by the vibratory motion of the cylinder is *not more simple* than the application of Shacklebars . . . from the cross of the Piston to the Crank."[25]

French continued to build engines in the West, and by 1830 Stacy Costell was building "vibratory" engines in Philadelphia. N. Rowland, also of Philadelphia,

Oscillating steam engine
of Ebenezer A. Lester, 1827.
Restored patent drawing in
records of the patent office,
National Archives,
Washington, D.C.

patented such an engine in 1833 and built at least one in that city. The Novelty Works of New York was building oscillating engines by 1833. In 1848 the statement was made that this type "is coming into general use in England, and a few have been erected here. It is then an object of some interest to our mechanics."[26] The pace of their acceptance accelerated during the 1850s, and by 1855 it could be said that "oscillating engines are very simple and compact, and are fast extending in use. We believe they have not yet been applied to locomotives," it was noted, "but they have to all other purposes."[27]

The chief advantage, as well as the major disadvantage, of the oscillating engine was that the motion of the cylinder regulated the flow of steam to the piston. It was explained in 1852 that the "oscillating cylinder is much employed for small land engines, as it affords a cheap substitute for the slide valve, if formed in the usual manner, by which one of the gudgeons or trunnions is made to regulate the entrance and exit of the steam, while the other gudgeon or trunnion can be employed to work the feed-pump."[28] The principal difficulty was that in order to prevent the escape of steam, the trunnions were fitted very tight, thus increasing friction which cut the power of the engine and caused undue wear on the trunnions themselves. Here too, a general improvement in strength and accuracy of construction helped to alleviate this problem.

The history of compound cylinder engines in this country is similar to that of other innovations. The engines were conceived in England, then either brought to America, or conceived independently in this country at an early date. After a period of uncertain success, they reached a fairly high level of development and use. Engines of two cylinders were made in England by Jonathan Hornblower in 1781 and Arthur Woolf in 1804. Their earliest use in the United States was in 1813, when Francis B. Ogden patented his compound engine. Using steam expansively, this engine was to work two pistons simultaneously at right angles to each other. It was explained that "one is at its maximum when the other is at its minimum, so that there is no necessity for a balance wheel as a reservoir of power: for when one piston ceases momentarily to act, the other instantaneously continues the motion in the same direction." It was pointed out that "there is nothing new in two cylinders, or in working by expansive steam, but the getting rid of the balance wheel by the simultaneous action of each piston in the manner above described . . . is new."[29]

Actually, Ogden's twin-cylinder engine was not of the classic compound type, developed by Woolf, since both cylinders were to use steam of the same pressure. In the true compound engine, the steam was introduced into the first cylinder under high pressure and then allowed to expand into the second cylinder under a lower pressure. This type seems not to have been used to any extent in America until the 1840s. Despite scattered reports of experiments with compound engines, it was estimated in 1856 that "this class of engine has received but partial trial in our country."[30]

The use of high-pressure steam expansively and the new positioning of cylinders were major innovations, but they do not account for the greatest part of the improvement of the steam engines in America. They were accompanied by an almost constant tinkering with the details of engine design: piston packing, valve action, governing, and linkage were the problems receiving particular attention.[31] Not all of this activity, of course, was productive of any real benefit to steam technology. Contemporary and recent literature are replete with references to the fact that the period following Watt's active work was one of attempting to absorb and convert his inventive genius into useful innovation.

A remark was made in 1819 that "on the expiration of Mr. Watt's patent, many ingenious mechanics attempted to improve the structure of the machine, and the records of the [British] patent office contain more upon this subject, than any other." As a result, "scarcely two following engines were made alike for many years, until, by the results of a vast deal of invention and experience, those methods . . . became settled into established forms; but none of them are superior to the original of Mr. Watt's."[32] Robert H. Thurston, a leading engineer, looking back over most of the nineteenth century, spoke of the period after 1850 as the "Period of Refinement" and said that "the inventor has been superseded by the engineer."[33] In other words, by midcentury the inventors had explored all the various alternatives inherent in the Watt engine—after that, improvement tended to be the province of the engineer with scientific training.

Because of the different needs of American engineers and the time lag in the exchange of technological information between this country and Great Britain, the major variations on Watt's basic engine design were not all adopted and perfected in the United States until about 1830. For example, it was not until 1827 that the basic and useful table "On the relative proportions of the various parts of the Boulton and Watt's, or low pressure, Steam Engine" was published in this country. This table, which showed "the very important results of the experience of Boulton and Watt, in proportioning the most essential parts of their steam engines," was said to have existed in manuscript form in all English engine shops. A copy was "procured, by an American engineer," who had recently visited England.[34]

After 1830, suggested improvements tended to be either repetitious, bizarre, or dealt with minor details of arrangement and construction. The best evidence of this leveling off of real improvement is the clear and dramatic rise in the number of attempts to perfect the rotary steam engine. This piece of mechanism, which absorbed so much inventive talent during the second quarter of the nineteenth century in America, had, like most other developments in steam engineering, been investigated by Watt in the eighteenth century. In this country, Oliver Evans had toyed with a rotary engine during the 1780s, but abandoned it for his high-pressure Columbian engine.

One early American rotary engine was built in Baltimore in 1819 by George Stiles, a former mayor of that city, and his son John S. Stiles. Three years before, Hezekiah Niles, ever ready to praise and encourage American ingenuity, announced with pride that a Mr. Curtis of Massachusetts had achieved "a desideratum in steam enginery, a complete rotary motion without a fly wheel or balance, an object of long and arduous research in Europe, of the greatest advantage to the mechanical world." The engine was reportedly "composed only of a cylinder containing a

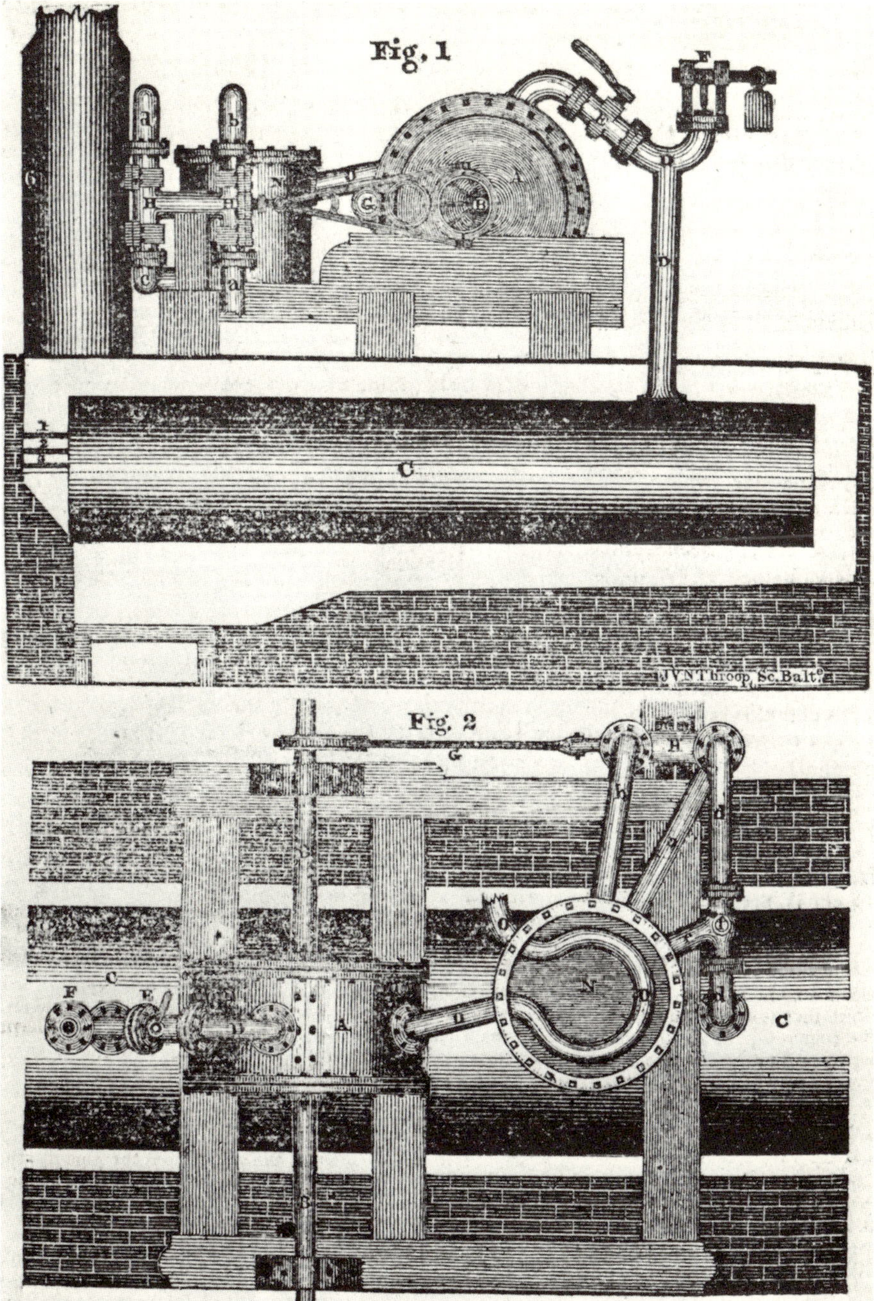

Rotary steam engine of George and John S. Stiles, 1819.
From *Niles' Weekly Register*, volume 17
(16 October 1819), page 114.

shaft-wheel with valves" and was touted as "vastly superior to those of Watt and Bolton, Evans, or any subsequent constructor." Curtis had one of his engines operating the sawmill of A. and N. Brown of Manhattan Island and had "after encountering a long series of difficulties . . . established a manufactory in Baltimore."[35]

The Stiles, father and son, were associated with Curtis in his Baltimore venture. A stationary engine used by the Stiles had been set at boring cannon during the War of 1812, then was used for ordinary shop work, and was finally used to operate a gristmill. After five years it was said to be "still working with equal power, without having received any material repairs." The four virtues claimed for it were its lightness, compactness, cheapness, and simplicity of motion. The last point was clinched, in Niles' mind, by the fact that the one then "working at Mr. Stiles' factory and mill is under the care of a black man, whose chief qualities, as needful to its management, are sobriety, and attention." Stiles, besides carrying on other types of work, made these rotary engines for general sale. Not untypically, in 1817 they even placed one of their engines aboard the boat *Surprize,* a steamboat which was said to be quite fast.[36]

Early in the nineteenth century there were sporadic reports of American mechanics, such as Stiles, having finally perfected a rotary engine, but it is significant that of the 123 patents issued in this country for rotary engines before 1860, only 20 were dated through 1830, while the remaining 103 were issued between 1831 and 1859. Plainly, interest increased rather than waned.[37]

The idea of the rotary engine was at least as old as the steam wheel of Hero of Alexandria, and apparently every inventor from the time of Watt had been struck with the notion that an engine which produced a rotary motion initially would be far superior to one in which the reciprocating motion had to be converted to circular. As late as 1829 such arbiters of American mechanical progress as Zachariah Allen were denouncing "the great loss of power resulting from the reciprocating movement of the piston of the common steam engines" and calling for an engine which would "apply the steam directly to turn a wheel, as water is applied to turn a water wheel, in order to produce at once a continued circular motion without the intervention of the beam, crank, and fly wheel."[38] But apparently a current of informed opinion was already flowing the other way. Remarking that "it may seem hardy to question a dogma, which has remained so long unquestioned," an inventor in 1822 endeavored to "show, contrary to the received theory, that there is no loss of power in the reciprocating movement of the common steam-engine."[39]

From the beginning of their publication, the *Franklin Journal* and the *Scientific American* launched vigorous campaigns against rotary engines. The editor of the former claimed to "have support of many practical engineers" for his opinion that "projectors are generally mistaken, respecting the loss of power, which they think, takes place in the crank."[40] The editor of the *Scientific American* hedged a bit by conceding that a successful rotary engine was theoretically possible, but pointed out that already, "the Mechanical ingenuity that has been expended in the invention and improvement of rotary engines, perhaps exceeds that of all other machines."[41] Between September 1848 and June 1849 he published a history of attempts to build such an engine (all abortive), with the avowed object "to throw light upon the subject, in order to stop many men of real original mechanical genius from wasting their time."[42] The editor claimed to have "seen no less than 60 representatives of different rotary engines" and blamed this continued activity on a "want of correct understanding of the principles of the steam engine."[43] The editor of still another technical journal professed to see "a perfect mania" for rotary engines in 1847.[44]

The Franklin Institute of Philadelphia had earlier been concerned with this same problem, remarking that "the great number of proposed improvements in the steam engine, equally manifest the possession of ingenuity, and the deficiency of knowledge and experience, in their projectors." To help remedy this tactfully described situation, the Institute appointed a committee on inventions to review and advise on improvements, offering its services gratis and promising to maintain "the utmost secrecy."[45] The persistence of interest in the rotary engine, however, cannot be explained solely on the basis of ignorance. The fact was that the common steam engine, as inherited from Watt and modified by Evans and others, appeared to inventors to have reached the practical limits of its perfection, and further major improvement was sought through a new departure.

Despite all counsel, efforts to develop the rotary engine continued. One persistent inventor was James A. Stewart of Cross Plains, Tennessee, who patented his rotary engine in 1841 and the following year had the firm of R. M. Roe and Company of New York City build a prototype. This engine was used for a short time in their shops, and Stewart was encouraged enough that he continued to try to perfect its details. About 1848 he made an engine of adequate power to operate a sawmill a few miles north of Nashville. The success of this engine led him to travel to St. Louis where the Eagle Foundry of Clark, Renfrew and Company manufactured these engines for general sale. The design of his engine was simple. Two cog wheels of long axis were enclosed in a tight-fitting iron jacket. Steam was introduced perpendicularly to the axes of the two wheels–the only moving parts–and along the line where they meshed. Except for Babbitt metal lining of the pillow blocks which carried the wheels,[46] the engine was built of iron throughout.

Despite persistent claims over a period of five years by Stewart and by his supporters that his engine had at last been perfected, it never was. Finally in 1853 he decided, as he said, "to abandon any further attempts to introduce my plan of rotary engine." He was prepared to admit that "an attempt to excel, materially, the reciprocating engine, in point of economy, by substituting a rotary, is chimerical." He persisted, nevertheless, in his belief that his principles were sound even though they failed in practice. The major faults of his engine were "its liability to derangement by expansion and contraction of metal, and the yielding or displacement of adjusting screws employed in maintaining the steam wheels in their proper position." Discouraged, Stewart moved to Texas and entered the business of flour milling.[47]

The efforts to build a better rotary engine, like most attempts to improve the steam engine, were guided by the supposed imperatives inherent within the engine itself. At least one major improvement in America, however, was imposed by

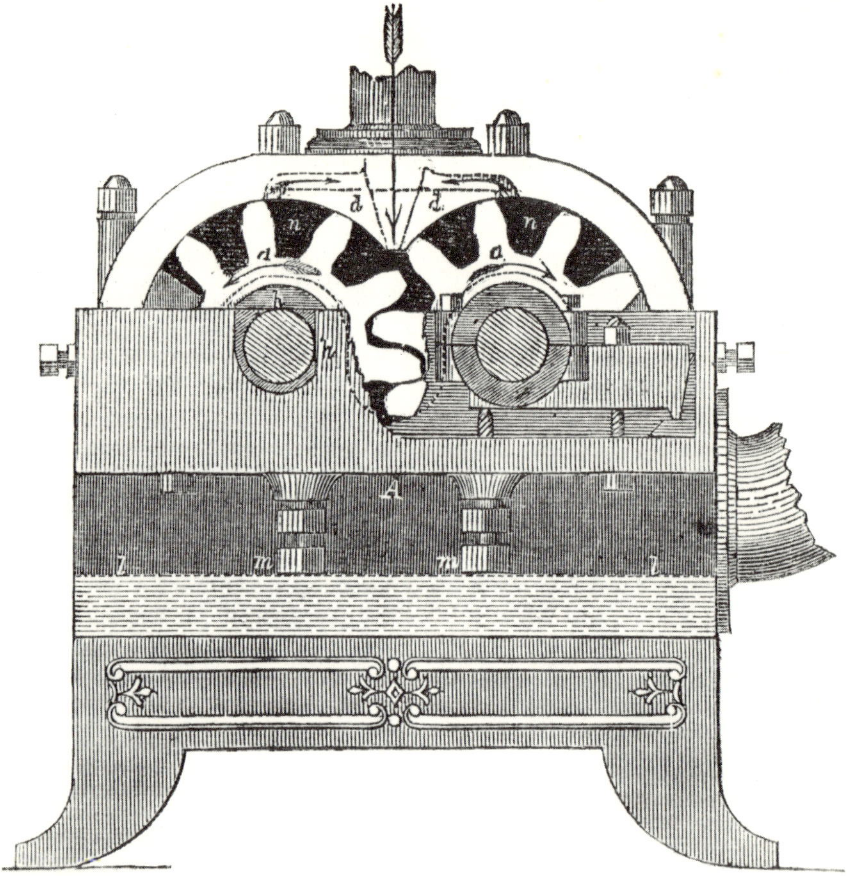

Rotary steam engine of James A. Stewart, 1850.
From *Scientific American*, volume 6
(9 November 1850), page 57.

philosophical and economic, rather than mechanical, desiderata. It was noted in 1857 that "within the last few years the attention of ingenious men in [Philadelphia] . . . , as well as in other places, has been directed largely to simplifying the Steam Engine, removing all essentially unnecessary parts, cheapening its price, and diminishing its size."[48] A part of this was merely the endless and undramatic battle to keep a piece of mechanism, once perfected, from being corrupted by innovation. The major incentive behind this activity, however, was the search for a smaller engine–perhaps even portable.

Thomas Jefferson had called for such engines as early as 1815. A small steam engine, he wrote, "applicable to our daily concerns, is infinitely more valuable than the greatest which can be used only for great objects." The former President was caught in the seeming contradiction between his enthusiasm for "improvement," and his revulsion at the idea of great industrial centers, based on steam-powered factories, springing up in America as they already had in England. His vision of small portable engines resolved the conflict, and he contemplated with satisfaction the idea that a small engine might be developed that would "raise from an adjacent well the water necessary for daily use; to wash the linen, knead the bread, beat the homony, churn the butter, turn the spit, and do all other household offices which require only a regular mechanical motion."[49] Thus the instrument of industrial vice in Europe would become the handmaiden of domestic virtue in America.

Jefferson was not alone in his hope that the steam engine might be domesticated, made democratic, and applied to the benefit of the common man. Hezekiah Niles quoted Oliver Evans to the effect that "the time would come when the powers of steam should be so well ascertained and so simply applied, that the very old women would do the common business of housewifery with it."[50] It was Evans' opinion, again according to Niles, that "the time would come, when, every person able to build a house costing 10,000 dollars, would provide himself with a steam engine, as a part of its furniture, to wash the clothes, scrub the floors, etc. etc."[51]

Writing in celebration of the national centennial in 1876, Abram Hewitt, a worthy representative of post-Civil War American civilization, announced that "if I were asked what elements had most to do with the swift progress of our country, I should answer, freedom and the steam engine. But deeper even than any organized declarations of outward forms of freedom lies the influence of the steam engine, which has been from the day of its birth, in spite of laws and dynasties, and all accidents of history, the great emancipator of man."[52] Beneath the thoroughly Victorian optimism and bravado of placing the steam engine above the Declaration of Independence and, presumably, above even the martyred Abraham Lincoln, that other "great emancipator of man," lay the intuitive realization, only recently rediscovered, that the unique contribution of American civilization to world progress lay in its standard of living for the common man.

It is not to be expected that the enginemaker, laboring to pare a few pounds or a troublesome lever from his steam engine, thought in terms of the message of America. But the very fact that this urge toward smaller and simpler engines was eminently practical was indicative of the thoroughness with which the concept of the mass market had already permeated the thinking of American industry. The *Scientific American* in 1853 stated that the want of a small portable engine is seriously felt by the public.

The farmer wants them to thresh his grain and cut his straw, to saw his wood, and as soon as they are properly constructed to draw his plow. The mechanic wants them for the various operations of his workshop, the manufacturer in a small way wants those that require but little room, and can be easily moved about as he may change his residence, and we may hope to see the day when they will be made so cheap that it will become almost a necessity of the household. The world is growing wiser and lazier every day.[53]

The advent of the internal combustion engine and the electric transmission of power removed some of the urgency from this quest, but during the first half of the nineteenth century, the attempt to provide a truly portable steam engine was a constant and important activity.

From the vantage point of his annual report for the year 1843, the Commissioner of Patents H. L. Ellsworth observed that "the great progress made in the steam-engine since Watt's time is more due to the increased knowledge of its properties, and better understanding of the general rules that should govern mechanical structures, and the improvement of all the mechanic arts, particularly in the working of iron, than to the improvements in the engine itself; although these are, by no means, insignificant."[54] This progress must be credited largely to the men who manufactured steam engines, aided by what one writer in Silliman's *American Journal of Science* referred to as "the patronage and encouragement of the public, and the auxiliary efforts of the Engineer and the Philosopher."[55]

The patronage and encouragement of the public is sufficiently documented by the wide and enthusiastic acceptance of steam power which characterized these years. Similarly, the multiplication and gradual professionalization of steam engineers need not be labored. Thomas Cooper, noting the activities of Robert Fulton, Oliver Evans, Benjamin Henry Latrobe, and others, wrote in 1814 that "so much exertion, and by such men, affords ground to hope that at no distant time, improvements may take place, that will put our Engineers of America, upon a par at least with those of England." The "three great requisites of a civil engineer," he declared, "are first a profound knowledge of mathematics, and an habitual facility of applying mathematical principles and calculations to mechanics and machinery. Secondly, a full knowledge of the modern science of chemistry. Thirdly, an habitual facility in drawing and designing with neatness and accuracy. To these should be added, actual observation and study of the principal machines in use and their application."[56] This ambitious prospectus remained desirable, rather than actual, for many years, despite suggestions that the federal government set up an academy to train steam engineers. The first engineers in this country were either self-tutored or trained in England, and frequently they were also manufacturers. Latrobe was one of the first and most successful among those who attempted to earn a living by the merit of his engineering and entrepreneurial talents alone.

Oliver Evans was a prime example of the inventor-manufacturer-engineer, who contributed so much to the early progress of steam engines in this country. Quoting the old saw that "truth lies in a well," Evans described the "most intense study" that preceded his new mechanisms. "After having a faint glimpse of the principle," he said, "it was with many toilsome and tedious steps that I attained a clear and distinct view. I received great assistance from the result of experiments made by others, which are to be found in scientific works."[57] He consulted the tables of John Dalton and speculated on the possibility of discovering a caloric-proof substance out of which boilers might be made.[58] So valuable did Evans find the available scientific works, that he suggested in order "to aid the progress of improvement in the arts and sciences," the federal government might, "at the expense of the community, employ ingenious persons, in every art and science, to make with care every experiment that might possibly lead to the extension of our knowledge of principles."[59] He specifically singled out investigations of solar radiation, presumably to raise steam in boilers, as an area in which the "expense of such investigations cannot, in many instances, be borne by those who have the mental powers to design." In such cases he thought "aid from government becomes necessary."[60]

Evans' views on the support of basic scientific research by the federal government were advanced, though shared by such important public figures as Thomas Jefferson and John Quincy Adams. In actual practice, engineers and manufacturers were forced to make use of what bits of scientific fact and encouragement they could find. They can be said to have been denied this resource only if the term science is limited in this case to a fully mature understanding of thermodynamics.[61] Even within this narrow definition, many early American scientists—notably Joseph Henry, Thomas Cooper, and James Renwick—concerned themselves with the theory and economy of steam. George Corliss, for example, was said to have had direct assistance in the invention of his Corliss engine from a professor of mathematics and natural philosophy at Brown University.[62]

It is probably a mistake, however, to so limit the definition of scientific aid to steam engineering. Manufacturers and projectors made constant recourse to whatever mathematical abilities they might have possessed. Furthermore, they sought the approval and endorsement of notable scientists for their improvements in engines. Such approbation was eagerly sought and widely advertised when received. Finally, they shared a common expectation that science would prove useful to the arts. There was near universal agreement with Alonzo Potter when he wrote that "on the one hand, science has furnished principles for the arts to apply: on the other, the arts have proposed problems for science to resolve; and this mutual aid and dependence have been the means of carrying both forward, at a rate continually accelerated."[63] It was admittedly a statement of faith—but, no doubt, it was a faith that sustained.

At the end of the American Revolution, the Watt steam engine presented American mechanics with a major and complex problem in technological assimilation. Faced with the staggering task of first learning to make and then to apply these engines, interested Americans made use of whatever English guidance was available through books and emigrant mechanics. Despite the handicap of having to work without a firm theoretical basis, these improvers applied to their task all the mathematics, experimental data, and knowledge of thermodynamics they could muster. In general, the first half of the nineteenth century was a period marked by consolidation and improvement rather than by dramatic breakthroughs in the technology of steam.

Experiments had been tried with several types of engines. Those best suited to steamboats, locomotives, and the various stationary purposes had become more or less standardized by midcentury. Improved design, more sturdy construction, and improved economy had also contributed to making steam power a widely accepted motive force. Simultaneously, a growing demand for engines had stimulated these same improvements so that steam technology had moved forward on a broad and connected front. Salients of more rapid development occurred, as in the case of the use of high-pressure steam, and in a few instances indifference or hostility brought lags in other areas. But on the whole, the advance was rapid and complete. Starting in the urban centers of the Middle Atlantic States, the new technology spread into the West, the South, and finally to New England. By midcentury the revolution was complete.

Notes

[1] *Scientific American,* volume 4 (5 May 1849), page 261.

[2] Evans, "Account," *Archives of Useful Knowledge,* volume 2 (1812), page 366; Evans, *Abortion,* pages 95-96.

[3] Evans, *Abortion,* p. 90n.

[4] Ibid., page 3.

[5] B. H. Latrobe to Charles T. Ingersoll, 17 January 1814, B. H. Latrobe MS. Coll., MHS.

[6] [Evans] Elisha, *Patent Right Oppression Exposed,* pages 63-64n.

[7] B. H. Latrobe to Andrew Hazelhurst, 19 October 1810, B. H. Latrobe MS. Coll., MHS.

[8] Daniel French to William Thornton, 12 February 1814, William Thornton papers, volume 4, LC.

[9] B. H. Latrobe to Robert Fulton, 1 March 1814, B. H. Latrobe MS. Coll., MHS.

[10] *Pittsburgh Gazette,* 9 April 1813.

[11] "Report of the Select Committee of the House of Commons, on Steam Navigation to India; with the Minutes of Evidence, Appendix, and Index. 1834," *Edinburgh Review,* volume 60 (January 1835), page 475n. This confirmed the statement of Frederick Graff, made in 1817, that Boulton and Watt engines were commonly worked from 10 to 20 psi. Frederick Graff to Robert Vaux, 9 July 1817, in *Communications . . . on the Explosions of Steam Boilers,* page 8.

[12] *American Railroad Journal,* volume 3 (27 December 1834), pages 806-807; Woodbury Report, passim.

[13] Evans, *Abortion,* pages 28-30.

[14] B. H. Latrobe to Thomas Cooper, 2 April 1814, B. H. Latrobe MS. Coll., MHS.

[15] *Ibid.*

[16] "Who First," *American Repertory,* volume 4 (1841), page 172.

[17] B. H. Latrobe to N. I. Roosevelt, 1 August 1815, B. H. Latrobe MS. Coll., MHS; B. H. Latrobe to F. B. Ogden, 23 January 1815, B. H. Latrobe MS. Coll., MHS.

[18] Quoted in "Who First," *American Repertory,* volume 4 (1841), page 172.

[19] Ibid., page 173.

[20] *Scientific American,* volume 10 (4 November 1854), page 61.

[21] John Fitch to Josiah Hornblower and Christopher Colles, 17 May 1786, Fitch papers, volume 1, LC.

[22] Josiah Hornblower to John Fitch, 24 May 1786, Fitch papers, volume 1, LC.

[23] Woodbury Report, page 156.

[24] *Scientific American,* volume 8 (18 December 1852), page 107. Seth Boyden is credited with developing "the cast-iron prome or bed." *Appletons' Cyclopaedia of American Biography,* volume 1, page 341.

[25] B. H. Latrobe to William Cooper, 14 October 1812, B. H. Latrobe MS. Coll., MHS.

[26] *Scientific American,* volume 3 (13 May 1848), page 265.

[27] Ibid., volume 10 (24 March 1855), page 218.

[28] Ibid., volume 8 (18 December 1852), page 107.

[29] [Cooper], "Account," *Emporium of Arts and Sciences,* volume 2 (1814), pages 473-474. J. Dowers, Jr., of Pittsburgh and Philadelphia, disputed some of the points of Ogden's patent. See B. H. Latrobe to Thomas Cooper, 2 April 1814, B. H. Latrobe MS. Coll., MHS. A Mr. Rodman of South Carolina, also planned a low-pressue, twin-cylinder engine in 1814. [Cooper], "Account," *Emporium of Arts and Sciences,* volume 2 (1814), page 373.

[30] *Scientific American,* volume 3 (13 May 1848), page 268; ibid., volume 11 (5 January 1856), page 129.

[31] The most famous and successful American innovation in the field during the middle years of the century was the valve gear of George Corliss. His impact was made in the period following that covered here, but something of his background will be found in Robert S. Holding, "George H. Corliss of Providence, Inventor," *Rhode Island History,* volume 5 (January 1946), pages 1-17.

[32] *Cyclopaedia; or, Universal Dictionary,* volume 34. For recent attempts to explain this in terms of "plateaus of technology," see Mumford, *Art and Technics,* page 84, and Milton Kerker, "Science and the Steam Engine," *Technology and Culture,* volume 2 (Fall, 1960), pages 385, 387.

[33] Thurston, *Growth of the Steam-Engine,* pages 304, 305. I would suggest that this period began about two decades earlier.

[34] *Franklin Journal,* volume 3 (May 1827), pages 336-339.

[35] *Niles' Weekly Register,* volume 10 (1 June 1816), page 219. Also on Curtis' engine see memo in hand of Nathan Read, 24 May 1816, and Bailey L. Pinsset [?] to William P. Page, 23 December 1816, Nathan Read papers, volumes 4 and 1, Essex Institute, Salem, Massachusetts.

[36] *Niles' Weekly Register,* volume 17 (16 October 1819), pages 97-98.

[37] *Index of Patents,* volume 3, pages 1219-1223. It should be noted that because of the quixotic nomenclature of the *Index,* one can never be sure of having counted all patents of any given type.

[38] Allen, *Science of Mechanics,* page 329.

[39] Minus Ward, "Ward's Alternating Steam-Engine," *American Journal of Science,* volume 4, no. 1 (1822), page 91. See also A. B. Quinby, "On Crank Motion," ibid., volume 7, no. 2 (1824), pages 316-323, and A. B. Quinby, "On Crank Motion, in reply to the remarks of the author of a Review in the North American," ibid., volume 9, no. 2 (1825), pages 317-324.

[40] *Franklin Journal,* volume 2 (December 1826), page 282.

[41] Scientific American, volume 2 (12 June 1847), page 301.

[42] Ibid., volume 4 (30 June 1849), page 325.

[43] Ibid., volume 3 (19 August 1848), page 380.

[44] *Eureka,* volume 2 (November 1847), page 44.

[45]*Franklin Journal,* volume 2 (September 1826), page 187.

[46]*Niles' Weekly Register,* volume 63 (19 November 1842), page 184; *Scientific American,* volume 3 (26 August 1848), page 390; ibid., volume 5 (27 July 1850), page 356; ibid., volume 6 (26 October 1850), page 43; ibid., volume 6 (16 November 1850), page 68; ibid., volume 8 (2 October 1852), page 21.

[47]*Scientific American,* volume 8 (2 April 1853), page 227.

[48]Freedley, *Philadelphia and its Manufactures,* page 316.

[49]Thomas Jefferson to George Fleming, 29 December 1815, in *Writings of Thomas Jefferson* (Washington), volume 6, page 505. Also see Edwin T. Martin, *Thomas Jefferson: Scientist* (New York, 1961), pages 72-73.

[50]Undated addenda to *Niles' Weekly Register,* volume 3 (September 1812-March 1813), page 1. It may be said that the vision of Jefferson and Evans has been fulfilled, in the sense that as late as 1945, two-thirds of the electricity in the United States was produced by steam power. Bureau of the Census, *Historical Statistics,* page 156. Cf. Mumford, *Art and Technics,* page 78.

[51]*Niles' Weekly Register,* volume 50 (16 April 1836), page 113.

[52]Abram S. Hewitt, "A Century of Mining and Metallurgy in the United States," *Transactions of the American Institute of Mining Engineers,* volume 5 (1876-1877), page 174.

[53]*Scientific American,* volume 9 (3 December 1853), page 89.

[54]*Report of the Commissioner of Patents, For the year 1843,* House Document 177, 28th Congress, 1st session, page 273.

[55]Quinby, "On Crank Motion," *American Journal of Science,* volume 9 (1825), pages 322-323.

[56][Cooper], "Account," *Emporium of Arts and Sciences,* volume 2 (1814), pages 374, 385.

[57]Evans, *Abortion,* p. 139.

[58][Evans], *Oliver Evans to His Counsel,* page 41; Evans, *Abortion,* page 134.

[59]Evans, *Abortion,* page 139.

[60]*Ibid.,* page 43.

[61]I have found, for example, no American references to the work of Sadi Carnot before the Civil War. The argument for scientific aid to the development of the steam engine during the 17th and 18th centuries is ably made in Milton Kerker, "Science and the Steam Engine," *Technology and Culture,* volume 2 (Fall, 1960), pages 381-390.

[62]Robert S. Holding, "George H. Corliss of Providence, Inventor," *Rhode Island History,* volume 5 (January 1946), page 7.

[63]Alonzo Potter, *The Principles of Science Applied to the Domestic and Mechanic Arts, and to Manufactures and Agriculture . . .* (Boston, 1841), pages 266-267.

Chapter 9

The Spreading Influence of Steam

Steam, therefore, until superceded by some more effective agent, will be the power principally relied upon to propel Machinery. . . .

EDWIN T. FREEDLEY (1859)[1]

Like most of the other elements of colonial American technology, the steam engine was brought to this country from England. It was a tool of severely limited usefulness and was not widely applied on this side of the Atlantic until James Watt made his improvements. Englishmen had been living in the colonies for well over a century when the first engine arrived in 1753. Anyone who was born in the colonies or anyone who had lived there for more than a few years had probably never seen such a machine, for they were as yet little used even in the mother country.

The improvements in steam technology made by Watt, on the eve of the American Revolution, greatly increased the efficiency and usefulness of the steam engine. He made it smaller, more powerful, and added immensely to its flexibility by giving it a rotative motion. Partly because of the revolutionary troubles already brewing, knowledge of these new developments was slow in reaching the colonies. Once independence had been won, Americans made themselves familiar with the new Watt engine and seized upon it as the solution to at least part of their economic problems.

Adequate transportation was the most pressing technological need of the new nation. Between the end of the Revolution and the turn of the century, more than a dozen inventors tried to find the proper combination of vessel, propelling mechanism, and steam engine that would produce a workable steamboat. All of them failed in their efforts, but most of these men then applied their new and hard-won knowledge of steam to stationary applications. The latter were less complex, and the Americans could draw upon already well-established English practice.

Engineers from England probably came to America after the war for the same reasons that attracted other Englishmen here. There were, however, added incentives: Boulton and Watt maintained a legal monopoly on engine manufacture in the old country; the export of steam engines from England was forbidden; and America was a virgin territory eager for the services of technical experts. Not all the names are recorded, but there must have been a good number of these engineers. John Fitch, Oliver Evans, Robert Fulton, John Stevens, and many other early American projectors had Englishmen in their employ or readily available for consultation. Benjamin Henry Latrobe was the most famous, and perhaps most ubiquitous, of these instructors, whose importance can hardly be overemphasized.

A few of the early steamboat experimenters turned from that activity to the building of stationary steam engines for general sale. About 1800, when the New Jersey Soho Works of Nicholas Roosevelt broke up, its trained workmen were scattered as far away as Pittsburgh. Several of these men taught engine builders of later importance or established engine shops of their own. When James Smallman and Oliver Evans set up shops in Philadelphia during the first decade of the nineteenth century, they made that city the first American center of steam-engine construction. As a result of the activities of Robert Fulton and the presence of former workmen of Roosevelt, New York City later became another center.

Many factors conspired to extend the building of steam engines to the West. The western waters were a continuing challenge to navigation; a lack of adequate waterpower for manufacturing, a rapidly growing population, a growing market for engines on the plantations of the South, and the dislocations of the War of 1812 all joined to make Pittsburgh a third center of engine building. From this location, activity spread to other cities along the Ohio River, and by the 1830s engine building was a major industry of the leading western cities.

By the end of the first third of the century, steam engines were in use in all parts of the United States and in every conceivable type of manufacturing establishment, from publishing houses to gunpowder mills. Released from the geographical imperatives of waterpower, mills began to appear in centers of industry, and American cities became important for manufacturing as well as commerce. While steam engines could be found in the heart of western forests and in sugarcane fields of the South, in 1838 it was discovered that most of them were located in urban centers. Here the power of engines was combined to run large mills, or subdivided to aid small shops. Both were essential to urban development.

Also by the 1830s steam engines were built in most of the cities of the country. Nearly any small town could produce a mechanic who could build or repair an occasional engine, but the large engine shops remained urban enterprises, producing engines as a part of a larger and more general ironwork business. In these shops, engines became standardized and their manufacture was rationalized. New York, Philadelphia, and Pittsburgh were still leading centers of engine building, but shops

Steam works of the Hayward Rubber Company, Colchester, Connecticut, 18__. The use of steam was often emphasized in advertisements, presumably to underscore the progressive nature of the firm's management and methods. In the building at the extreme left, note the man cutting wood for the furnace.

were also located in New England and the South. The business of making steam engines had advanced so far that it was probably no more speculative or hazardous than any other type of manufacture. Engines built in America were even beginning to be exported.

From the beginning American engineers sought to improve upon and to advance the uses of the English steam engines. The main concern, however, was to assimilate those English improvements which appeared during the last quarter of the eighteenth century—it having taken engineers in England that long to convert Watt's inventions and insights into practical use. American engineers, except in the field of high-pressure steam, ran some years behind their English counterparts, but by the 1830s they had experimented with and chosen between all the English developments. By midcentury a standard American stationary engine was evolved, and further improvements were sought in the direction of a workable rotary engine.

The results of the power revolution are as obvious as they are difficult to document. Indices of coal and iron production, for example, show that these basic industries experienced a phenomenal growth during the antebellum years. The introduction of steam engines no doubt accounts for a good deal of this activity. Eighteenth-century American technology was largely based upon the use of wood. Steam engines were the first large machines which could be made only of iron, and in the years before there was any demand for railroad or structural iron, they led in stimulating iron production.

Of equal or greater significance was the liberating influence of steam. The harnessing of water—through the use of power machinery and in canal transportation—had begun the transformation of America into a unified and industrial nation by the end of the eighteenth century. But water was not universally available. The steam engine, however, provided any amount of power at every site; it released industry from the geographic shackles of its previous reliance upon waterpower streams. The industrial pattern of New Castle County, Delaware, is an excellent case in point.

Situated on the Fall Line, where the swift-running Brandywine River flows into the navigable tidewater of the Delaware Bay, the village of Wilmington developed a flour-milling industry that was famous even in Europe during the latter half of the eighteenth century. As the industry of America diversified, and iron, textiles, gunpowder, and a host of other products were added to the manufactures of the area, the newly built mills could find waterpower space only by moving farther upstream along the Brandywine. By the 1820s an almost solid line of mills crowded the banks of the river well into Pennsylvania, and establishments had spilled over onto the smaller streams tributary to the Christiana. A prospective miller had only a small chance of buying and converting a flour mill in Wilmington: his only other choice was to move ten or twenty miles from the markets and transportation facilities of the town. After the advent of steam, millers no longer had to cling to the Fall Line or move upstream into the Piedmont.

It was now possible to build along the broad waterfront or along the railroad spurs, where transportation facilities were adequate. By 1857 at least forty-nine engines, ranging in power from one horsepower to two hundred, were operating in the town of Wilmington, driving cotton factories and tanneries, iron-rolling mills, and even gristmills. The warning of a local newspaper editor only three years before, "that Wilmington must, if she would prosper rapidly, become a manufacturing city," was already being heeded, and steam engines were becoming the instrument of the city's new prosperity.[2]

Before the use of steam engines, mills were limited not only geographically but also in size. By mid-nineteenth century the use of efficient water turbines and the harnessing of large streams, especially in New England, had allowed some mills to expand into considerable establishments. For the vast majority of waterpower sites, however, maximum available horsepower continued to be small. Once again the Wilmington area furnished a typical example. In 1857 about sixteen hundred horsepower was developed by steam in Wilmington. At the same time, the nearby Red Clay Creek had been supporting a dozen mills since the eighteenth century on a developed fraction of that stream's potential seven hundred horsepower.[3] The inevitable result of this kind of power differential was that most manufacturers expanded their facilities through the use of steam power. A good water turbine could extract upward of 90 percent of the work available in a given volume and fall of water. But beyond that there was no hope of further power.

The use of steam not only raised the upper limits of available power dramatically, but also lowered the minimum. Single horsepower steam engines were readily available at a price that a small manufacturer could afford. Such a small shop could now operate in the heart of a large city, where waterpower had never existed. As early as 1834 New York City contained seventy-six steam engines, developing a total of 858 horsepower, or an average of only a little over 11 horsepower each.[4] Nearly all of this represented a net gain of power available for manufacturing purposes.

The removal of limitations on location and power was only the most obvious boon offered to American industry by the introduction of steam power. The very demand for engines had broad implications for the whole metal-trade industry. The need for boiler plate stimulated the building of rolling mills when there was little other use for sheet iron, and boilermakers were the pioneers in the rise of the iron shipbuilding industry. The call for ever larger castings and the need for larger and accurate boring machinery forced a rapid development in these areas. In turn, the spiraling demand for more sizable iron products stimulated furnace masters to erect steam engines to power their greater blast machines. Since engine components were mainly cast, rather than wrought, their multiple production led inevitably to a system of uniform parts. Here was no flash of genius—but the logical and probably unconscious result of sound business practices. Coal was ripped from the earth to aid in building a hoard of new engines which in turn demanded an increasing amount of the same fuel. The beginnings of steam power were not only dramatic but also full of promise for the future.

Notes

[1] Freedley, *Philadelphia and Its Manufactures,* page 36.

[2] *Delaware State Journal* (Wilmington), 11 August 1857; *Delaware Gazette* (Wilmington), 10 January 1854.

[3] *Delaware State Journal,* 11 August 1857; *Wilmington Daily Commercial,* 10 September 1869.

[4] *American Railroad Journal,* volume 4 (8 August 1835), page 258.

Bibliography

Unpublished Manuscripts

Baker Library, Harvard University Graduate School of Business Administration, Cambridge, Massachusetts
Samuel Slater collection

Essex Institute, Salem, Massachusetts
Nathan Read papers

The Filson Club, Louisville, Kentucky
Various collections

Franklin Institute, Philadelphia, Pennsylvania
Oliver Evans papers

Library of Congress, Washington, D.C.
Isaac Briggs papers
John Fitch papers
William Thornton papers

Maryland Historical Society, Baltimore, Maryland

National Archives, Washington, D.C.
Boulton correspondence, record group 104, records of the Bureau of the Mint
Census of manufactures, schedules of the, record group 29, records of the Bureau of the Census
Patent drawings, restored, record group 241, records of the Patent Office
State, miscellaneous letters of the Department of, record group 59, records of the Department of State

New-York Historical Society, New York, New York
Robert Fulton papers

New York Public Library, New York, New York
Robert Fulton papers, Montague collection

Pennsylvania, Historical Society of, Philadelphia, Pennsylvania
Various collections

Washington, University of, Seattle, Washington
Manning Ferguson Force, papers collected by Peter Force re John Fitch, manuscript collection

Periodicals

American Institute of Mining Engineers, *Transactions,* 72 volumes, 1871-1925.

The American Journal of Improvements in the Useful Arts, and Mirror of the Patent Office in the United States (Washington, D.C.), 1 volume, 1828.

The American Journal of Science and Arts, 80 volumes, 1818-1860.

American Mechanics' Magazine (New York), 2 volumes, 1825-1826.

The American Repertory of Arts, Sciences, and Manufactures (New York), 4 volumes, 1840-1842.

American Society of Mechanical Engineers, *Transactions,* 48 volumes, 1880-1926.

The American Weekly Messenger; or, Register of State Papers, History and Politics. For 1813-14 (Philadelphia), 1 volume, 1814.

Archives of Useful Knowledge (Philadelphia), 3 volumes, 1810-1813.

Emporium of Arts and Sciences (Philadelphia), old series, 2 volumes, 1812-1813; new series, 3 volumes, 1813-1814.

Eureka; or, The Journal of the National Association of Inventors (New York), 2 volumes, 1847-1848.

Hunt's Merchants' Magazine and Commercial Review (New York), 58 volumes, 1839-1868.

The Medical Repository (New York), 11 volumes, 1797-1808.

The National Magazine and Industrial Record (New York), 2 volumes, 1845-1846.

Niles' Weekly Register (Baltimore), 1811-1849.

Quarterly Journal of Agriculture, Mechanics, and Manufactures (New York), 1 volume, 1834.

The Universal Asylum, and Columbian Magazine (Philadelphia), 9 volumes, 1787-1792.

The Useful Cabinet (Boston), 1 volume, 1808.

Books and Articles

[Adams, John]. *Diary and Autobiography of John Adams,* edited by L. H. Butterfield. 4 volumes, Cambridge, 1961.

A Digest of Patents, Issued by the United States, from 1790 to January 1, 1839 . . . . Washington, D.C., 1840.

Albion, Robert Greenhalgh. *The Rise of New York Port, 1815-1860.* New York, 1939.

Allen, Zachariah. *The Science of Mechanics, as applied to the Present Improvements in Europe, and in the United States of America: Adapted as a Manual for Mechanics and Manufacturers.* Providence, 1829.

An Alphabetical List of All Patents and Patentees on Record in the Patent Office. [Washington, D.C., 1830?]

"An Old Grasshopper Engine," *Iron Age,* volume 53 (25 January 1894), pages 151-152.

Appletons' Cyclopaedia of American Biography, edited by James Grant Wilson and John Fiske. 6 volumes, New York, 1888-1889.

Arndt, Karl J. "Three Hungarian Travelers Visit Economy," *Pennsylvania Magazine of History and Biography,* volume 79 (April 1955), pages 197-216.

Ashton, T. S. *The Industrial Revolution, 1760-1830.* London, 1948.

________. *Iron and Steel in the Industrial Revolution.* Manchester, 1924.

Babcock, George H. "Substitutes for Steam," *Transactions of the American Society of Mechanical Engineers,* volume 7 (1885-1886), pages 680-741.

Bancroft, Joseph. *Census of the City of Savannah, together with Statistics Relating to the Trade, Commerce, Mechanical Arts...* 2d edition, Savannah, 1848.

[Bangs, Isaac]. "New York in 1776," *Historical Magazine,* volume 4, 2d series (December 1868), pages 305-306.

[Batchelder, Samuel]. *Introduction and Early Progress of the Cotton Manufacture in the United States.* Boston, 1863.

Bathe, Greville. *An Engineers's Miscellany.* Philadelphia, 1938.

________ and Dorothy Bathe. *Jacob Perkins: His Inventions, His Times, and His Contemporaries.* Philadelphia, 1943.

_______. *Oliver Evans: A Chronicle of Early American Engineering.* Philadelphia, 1935.

Bigelow, John P., compiler. *Statistical Tables: Exhibiting the Condition and Products of Certain Branches of Industry in Massachusetts, for the Year Ending April 1, 1837.* Boston, 1838.

Binder, Frederick M. "Pennsylvania Coal and the Beginnings of American Steam Navigation," *Pennsylvania Magazine of History and Biography,* volume 83 (October 1959), pages 420-445.

Bining, Arthur Cecil. *Pennsylvania Iron Manufacture in the Eighteenth Century.* Harrisburg, 1938.

"Biographical Sketch of William Henry of Lancaster County, Pennsylvania," *Pennsylvania Magazine of History and Biography,* volume 27 (1903), pages 91-93.

Bishop, J. Leander. *A History of American Manufactures from 1608 to 1860.* 3 volumes, Philadelphia, 1861-1868.

Blake, Nelson Manfred. *Water for the Cities: A History of The Urban Water Supply Problem In the United States.* Syracuse, 1956.

Bowden, Witt. *Industrial Society in England Towards the End of the Eighteenth Century.* 2d edition, New York, 1965.

Brainard, Newton C. "Apollos Kinsley," *Connecticut Historical Society Bulletin,* volume 26 (January 1961), pages 12-20.

Bridenbaugh, Carl. *The Colonial Craftsman.* New York, 1950.

Brown, James S. *Allaire's Lost Empire: A Story of the Forges and Furnaces of the Manasquam.* Freehold, New Jersey, 1958.

Bruchey, Stuart. *The Roots of American Economic Growth, 1607-1861: An Essay in Social Causation.* New York, 1965.

Buck, Solon J., and Elizabeth Hawthorn Buck. *The Planting of Civilization in Western Pennsylvania.* Pittsburgh, 1939.

Calendar of the Miscellaneous Letters Received by the Department of State. From the Organization of the Government to 1820. Washington, D.C., 1897.

Calhoun, Daniel Hovey. *The American Civil Engineer: Origins and Conflict.* Cambridge, 1960.

Calvert, Monte A. *The Mechanical Engineer in America, 1830-1910: Professional Cultures in Conflict.* Baltimore, 1967.

Carnot, N.-L.-S. *Reflections on the Motive Power of Heat,* edited by R.H. Thurston. 2d revised edition, New York, 1897.

[Chamberlain, E.] *The Indiana Gazetter, or Topographical Dictionary of the State of Indiana.* Indianapolis, 1849.

"Charles T. James, of Rhode Island," *De Bow's Southern and Western Review,* volume 9 (December 1850), pages 671-675.

Cist, Charles. *Cincinnati in 1841: Its Early Annals and Future Prospects.* Cincinnati, 1841.

Clark, Victor S. *History of Manufactures in the United States.* 3 volumes, New York, 1929.

Clowes, John. "Steam and Water as Motive Powers," *National Magazine and Industrial Record,* volume 1 (October 1845), pages 443-448.

Colden, Cadwallader D. *The Life of Robert Fulton.* New York, 1817.

Coleman, J. Winston, Jr. *Steamboats on the Kentucky River.* Lexington, 1960.

Collins, Richard H., editor *Collins' Historical Sketches of Kentucky. History of Kentucky: By the Late Lewis Collins... Revised, Enlarged Three-Fold, and Brought Down to the Year 1874.* Louisville, 1877.

Communications Received by the Committee of the Franklin Institute, on the Explosions of Steam Boilers.... Philadelphia, 1832.

[Convention of Iron Masters]. *Documents Relating to the Manufacture of Iron in Pennsylvania.* Philadelphia, 1850.

[Cooper, Thomas]. "An Account of the Steam Engine," *Emporium of Arts and Sciences,* volume 2, new series (December 1813), pages 1-119; (February 1814), pages 169-220; (April 1814), pages 333-385.

Copeland, Melvin Thomas. *The Cotton Manufacturing Industry of the United States.* Cambridge, 1912.

Coxe, Tench. *An Address to an Assembly of the Friends of American Manufactures....* Philadelphia, 1787.

Curse, Englehart. *The Projector Detected, or, Some Strictures, On the Plan of Mr. James Rumsey's Steam Boat.* Baltimore, 1788.

Cutler, William Parker and Julia Perkins Cutler, editors. *Life, Journals, and Correspondence of Rev. Manasseh Cutler, LL.D.* 2 volumes, Cincinnati, 1888.

Dangerfield, George. *Chancellor Robert R. Livingston of New York, 1746-1813.* New York, 1960.

Davis, Madison. "The Old Cannon Foundry above Georgetown, D.C., and Its First Owner, Henry Foxall," *Records of the Columbia Historical Society, Washington, D.C.* volume 11 (1908), pages 16-70.

Dawson, J.L., and H.W. DeSaussure. *Census of the City of Charleston, South Carolina, for the Year 1848...* Charleston, 1849.

[Deane, Silas]. *Correspondence between Silas Deane and His Brothers and Their Business and Political Associates, 1771-1795,* Collections of the Connecticut Historical Society, volume 23. Hartford, 1930.

Department for Supplying the City with Water. *Annual Report of the Chief Engineer of the Water Department of the City of Philadelphia For the Year 1875.* Philadelphia, 1876.

DeWitt, Francis, compiler. *Statistical Information Relating to Certain Branches of Industry in Massachusetts, For the Year Ending June 1, 1855.* Boston, 1856.

Deyrup, Felicia Johnson. *Arms Makers of the Connecticut Valley: A Regional Study of the Economic Development of the Small Arms Industry, 1798-1870.* Smith College Studies in History, volume 33. Northampton, 1948.

Dickinson, H.W. *A Short History of the Steam Engine.* Cambridge, 1939.

________. *Robert Fulton, Engineer and Artist: His Life and Works.* London, 1913.

________. "The Steam-Engine to 1830," *A History of Technology,* edited by Charles Singer et al., volume 4 (New York, 1958), pages 168-198.

Dictionary of American Biography, edited by Allen Johnson and Dumas Malone. 22 volumes, New York, 1928-1944.

[Dix, John Adams]. *Sketch of the Resources of the City of New-York...* New-York, 1827.

Doolittle, Isaac. "Remarks on the Revolving Steam Engine of Morey," *American Journal of Science,* volume 2 (April 1820), pages 101-105.

Dow, Charles H. *History of Steam Navigation between New York and Providence.* New York, 1877.

Drake, B., and E.D. Mansfield. *Cincinnati in 1826.* Cincinnati, 1827.

Drake, Daniel. *Natural and Statistical View, or Picture of Cincinnati and the Miami Country....* Cincinnati, 1815.

Dugan, Frances L.S., and Jacqueline P. Bull, editors. *Bluegrass Craftsman: Being the Reminiscences of Ebenezer Hiram Stedman, Papermaker 1808-1885.* Louisville, 1959.

Dupree, A. Hunter. *Science in the Federal Government: A History of Policies and Activities to 1940.* Cambridge, 1957.

Durand, William Frederick. *Robert Henry Thurston: A Biography. The Record of a Life of Achievement as Engineer, Educator, and Author.* New York, 1929.

Durfee, W.F. "Some Ancient Reaction Engines," *Scientific American,* volume 70 (17 March 1894), pages 165-166.

Eaton, Amos. *Prodromus of A Practical Treatise on the Mathematical Arts: Containing Directions for Surveying and Engineering.* Troy, New York, 1838.

Eavenson, Howard N. *The First Century and a Quarter of American Coal Industry.* Pittsburgh, 1942.

________. *The Pittsburgh Coal Bed: Its Early History and Development.* New York, 1938.

"Ericsson's Caloric Engine," *Hunt's Merchants' Magazine,* volume 27 (July 1852), pages 19-28.

Evans, Oliver. *The Abortion of the Young Steam Engineer's Guide.* Philadelphia, 1805.

________. "Account of the Steam Engines of Oliver Evans," *Archives of Useful Knowledge,* volume 2 (April 1812), pages 362-369.

________. *Oliver Evans to His Counsel, Who Are Engaged in Defense of His Patent Rights.* No place, [1816].

[Evans, Oliver]. Patrick N.I. Elisha. *Patent Right Oppression Exposed; or, Knavery Detected.* Philadelphia, 1813.

Farey, John. *A Treatise on the Steam Engine, Historical, Practical, and Descriptive.* London, 1827.

Farrell, Gabriel, Jr. *Capt. Samuel Morey; Who Built a Steamboat Fourteen Years Before Fulton.* Manchester, New Hampshire, 1915.

Ferguson, Eugene S., editor. *Early Engineering Reminiscenses (1815-40) of George Escol Sellers.* U.S. National Museum Bulletin 238. Washington, D.C. 1965.

________. "John Ericsson and the Age of Caloric," U.S. National Museum Bulletin 228 (Washington, D.C., 1961), pages 41-60.

Fitch, John. *The Original Steam-Boat Supported: or, A Reply to Mr. James Rumsey's Pamphlet.* Philadelphia, 1788.

Flexner, James Thomas. *Steamboats Come True: American Inventors in Action.* New York, 1944.

Franklin, Benjamin. *The Complete Works of Benjamin Franklin,* edited by John Bigelow. 10 volumes, New York, 1887-1888.

Fraser, Charles. "The Moral Influence of Steam," *Hunt's Merchants' Magazine,* volume 14 (June 1846), pages 499-515.

Freedley, Edwin T. *Philadelphia and its Manufactures: A Hand-Book Exhibiting the Development, Variety, and Statistics of the Manufacturing Industry of Philadelphia in 1857.* Philadelphia, 1859.

French, B.F. *History of the Rise and Progress of the Iron Trade of the*

United States, from 1621 to 1857. New York, 1858.

[Friends of Domestic Industry]. *Journal of the Proceedings of the Friends of Domestic Industry, in General Convention Met at the City of New York, October 26, 1831.* Baltimore, 1831.

Fritz, John. *Autobiography of John Fritz.* New York, 1912.

Fuller, Grace Pierpont. *An Introduction to the History of Connecticut as a Manufacturing State.* Smith College Studies in History, volume 1. Northampton, 1915.

Galloway, Elijah. *History and Progress of the Steam Engine....* London, 1831.

Goodale, Stephen L. *Chronology of Iron and Steel.* Pittsburgh, 1920.

Gordon, Thomas F. *A Gazetteer of the State of New Jersey...* Trenton, 1834.

[Gottesman, Rita Susswein]. *The Arts and Crafts in New York, 1726-1776; Advertisements and News Items from New York City Newspapers.* New York, 1938.

————. *The Arts and Crafts in New York, 1777-1799.* New York, 1954.

Graff, Frederick. "Notice of the Earliest Steam Engines used in the United States," *Journal of the Franklin Institute,* volume 55 (April 1853), pages 269-271.

Grammer, John, Jr. "Account of the Coal Mines in the Vicinity of Richmond, Virginia," *American Journal of Science,* volume 1 (1818), pages 125-130.

Great Britain, Parliament: House of Commons. Sessional Papers (1825), volume 5. *Report of the Select Committee on the Laws Relating to the Export of Tools and Machinery.*

————. Sessional Papers (1841), volume 7. *Reports from Committees... Exportation of Machinery.*

Greeley, Horace. *Art and Industry as Represented in the Exhibition at the Crystal Palace New York--1853-4...* New York, 1853.

Gregg, William. "Domestic Industry--Manufactures at the South," [De Bow's] *Commercial Review,* volume 8 (February 1850), pages 134-146.

Griffin, Richard W., and Diffee W. Standard. "The Cotton Textile Industry in Ante-Bellum North Carolina. Part II: An era of Boom and Consolidation, 1830-1860," *North Carolina Historical Review,* volume 34 (April 1957), pages 131-164.

Griffin, Richard W. "Manufacturing Interests of Mississippi Planters, 1810-1832," *Journal of Mississippi History,* volume 22 (April 1960), pages 110-122.

————. "Pro-Industrial Sentiment and Cotton Factories in Arkansas, 1820-1863," *Arkansas Historical Quarterly,* volume 15 (Summer, 1956), pages 125-139.

Hamilton, Alexander. *Alexander Hamilton's Papers on Public Credit, Commerce, and Finance,* edited by Samuel McKee, Jr. New York, 1957.

Hamlin, Talbot. *Benjamin Henry Latrobe.* New York, 1955.

Harrison, Joseph, Jr. *The Locomotive Engine,and Philadelphia's Share in Its Early Improvements.* Revised edition, Philadelphia, 1872.

Haswell, Charles H. "Early Marine Engineering in the United States," *Engineering News,* volume 39 (19 May 1898), pages 315-316.

————. "Reminiscences of Early Marine Steam Engine Construction and Steam Navigation in the United States of America from 1807 to 1850," *Transactions of the Institution of Naval Architects,* volume 40 (London, 1898), pages 104-110.

———. *Reminiscences of an Octogenarian of the City of New York (1816 to 1860).* New York, 1896.

Hedges, Isaac A. *Sugar Canes and Their Products Culture and Manufacture.* St. Louis, 1879.

Hedges, James B. *The Browns of Providence Plantations: Colonial Years.* Cambridge, 1952.

Heirs of Robert Fulton, 27 January 1846. House Document 145, 29th Congress, 1st session.

Henry M. Shreve–Snag Boat, 28 February 1843. House Document 272, 27th Congress, 3d session.

Hermelin, Samuel Gustaf. *Report About the Mines in the United States of America, 1783,* translated by Amandus Johnson. Philadelphia, 1931.

Hewitt, Abram S. "A Century of Mining and Metallurgy in the United States," *Transactions of the American Institute of Mining Engineers,* volume 5 (1876-1877), pages 164-196.

———. "Our Founder," *The Stevens Indicator,* volume 14 (April 1897), pages 107-121.

Hibben, Henry B. *Navy-Yard, Washington. History from Organization 1799 to Present Date.* Senate Executive Document 22, 51st Congress, 1st session. Washington, D.C. 1890.

Hildreth, S.P. "Observations on the Bituminous Coal deposits of the valley of the Ohio...," *American Journal of Science,* volume 29 (January 1836), pages 1-154.

Hinchman, Walter S. *Holmes Hinkley: An Industrial Pioneer, 1793-1866.* Cambridge, 1913.

"History of the First Steam-Ship that Crossed the Atlantic," *Hunt's Merchants' Magazine,* volume 23 (December 1850), page 693.

"The History of the Steam Engine in America," *Journal of the Franklin Institute,* volume 102 (October 1876), pages 253-268.

Hodge, P.R. *The Steam Engine, Its Origin and Gradual Improvement, from the Time of Hero to the Present Day; as Adapted to Manufactures, Locomotion and Navigation.* New York, 1840.

Holding, Robert S. "George H. Corliss of Providence, Inventor," *Rhode Island History,* volume 5 (January 1946), pages 1-17.

Hoopes, Penrose Robinson. *Connecticut's Contribution to the Development of the Steamboat.* Committee on Historical Publications, volume 53. New Haven, 1936.

Howe, Henry. *Memoirs of the Most Eminent American Mechanics: Also, Lives of Distinguished European Mechanics; together with Anecdotes, Descriptions, etc. etc. Relating to the Mechanic Arts.* New York, 1842.

Hunter, Louis C. "The Invention of the Western Steamboat," *Journal of Economic History,* volume 3 (November 1943), pages 201-220.

———. *Steamboats on the Western Rivers: An Economic and Technological History.* Cambridge, 1949.

———. *Studies in the Economic History of the Ohio Valley: Seasonal Aspects of Industry and Commerce Before the Age of Big Business [and] The Beginnings of Industrial Combination.* Smith College Studies in History, volume 19. Northampton, 1933-1934.

Hussey, Miriam. *From Merchants to "Colour Men"; Five Generations of Samuel Wetherill's White Lead Business.* Philadelphia, 1956.

Hutton, F. R. "First Stationary Steam Engines in America," *Transactions of the American Society of Mechanical Engineers,* volume 15 (1894), pages 982-997.

Jackson, Joseph. *Early Philadelphia Architects and Engineers.* Philadelphia, 1923.

[Jefferson, Thomas]. *The Papers of Thomas Jefferson,* edited by Julian P. Boyd. Princeton, 1950-.

________. *The Writings of Thomas Jefferson,* edited by Paul Leicester Ford. 10 volumes, New York, 1892-1899.

________. *The Writings of Thomas Jefferson,* edited by H. A. Washington. 9 volumes, Washington, D.C., 1853-1854.

John Borden Fitch et al., 20 January 1846. House Document 124, 29th Congress, 1st session.

Jones, S[amuel]. *Pittsburgh in the Year 1826 . . . .* Pittsburgh, 1826.

Jordan, Francis, Jr. *The Life of William Henry of Lancaster, Pennsylvania, 1729-1786, Patriot, Military Officer, Inventor of the Steamboat.* Lancaster, 1910.

Jordan, John W. "Moravian Immigration to Pennsylvania, 1734-1765," *Pennsylvania Magazine of History and Biography,* volume 33 (1909), pages 228-248.

Kerker, Milton. "Sadi Carnot and the Steam Engine Engineers," *ISIS,* volume 51 (September 1960), pages 257-270.

________. "Science and the Steam Engine," *Technology and Culture,* volume 2 (Fall, 1960), pages 381-390.

Kirby, Richard Shelton, editor. *Inventors and Engineers of Old New Haven.* New Haven, 1939.

Kuhlmann, Charles Byron. *The Development of the Flour-Milling Industry in the United States: With Special Reference to the Industry in Minneapolis.* Boston, 1929.

Laidley, W. S. *History of Charleston and Kanawha County West Virginia and Representative Citizens.* Chicago, [1911].

Laing, James T. "The Early Development of the Coal Industry in the Western Counties Of Virginia, 1800-1865," *West Virginia History,* volume 27 (January 1966), pages 144-155.

Lander, Ernest M., Jr. "Charleston: Manufacturing Center of the Old South," *Journal of Southern History,* volume 26, (August 1960), pages 330-351.

________. "The Iron Industry in Ante-Bellum South Carolina," *Journal of Southern History,* volume 20 (August 1954), pages 337-355.

La Rochefoucault Liancourt, Duke de. *Travels through the United States of North America, the Country of the Iroquois, and Upper Canada, in the Years 1795, 1796, and 1797 . . . .* 2 volumes, London, 1799.

Lathrop, William G. *The Brass Industry in Connecticut: A Study of the Origin and the Development of the Brass Industry in the Naugatuck Valley.* Shelton, Connecticut, 1909.

Latrobe, B. H. *American Copper Mines.* No place or date.

________. *An Answer to the Joint Committee of the Select and Common Councils of Philadelphia, on the Subject of a Plan for Supplying the City with Water, &c.* No place, 2 March 1799.

________. "First Report of Benjamin Henry Latrobe, to the American Philosophical Society, held at Philadelphia; in answer to the enquirey of the Society of Rotterdam, Whether any, and what improvements have been made in the construction of Steam-Engines in America?", *Transactions of the American Philosophical Society,* volume 6, part 1 (1804), pages 89-98.

[Latrobe, Benjamin Henry]. *View of the Practicability and Means of Supplying the City of Philadelphia with Wholesome Water. In a Letter to John Miller, Esquire, from B. Henry Latrobe, Engineer. December 29th 1798.* Philadelphia, 1799.

Lesley, J. P. *The Iron Manufacturer's Guide to the Furnaces, Forges and Rolling Mills of the United States* . . . . New York, 1859.

Letter from the Secretary of State accompanied with a list of the names of persons who have invented any new and useful art, machine, manufacture or composition of matter, or any improvement thereon, and to whom patents have been issued . . . . Washington, D.C., 1805.

Lincoln, Jonathan Thayer. "The Beginnings of the Machine Age in New England: David Wilkinson of Pawtucket," *New England Quarterly*, volume 6 (December 1933), pages 716-732.

Lippincott, Isaac. *A History of Manufactures in the Ohio Valley to the Year 1860.* No place, 1914.

Lord, John. *Capital and Steam-Power, 1750-1800.* London, 1923.

Lord, John C. "Samuel Wilkeson," *Publications of the Buffalo Historical Society*, volume 4 (1896), pages 71-85.

Loree, L. F. *First Steam ("Fire") Engine in America.* No place or date.

Lyford, W. G. *The Western Address Directory . . . (For the Year 1837)* . . . . Baltimore, 1837.

Macfarlane, John J. *Manufacturing in Philadelphia, 1683-1912.* Philadelphia, 1912.

Macfarlane, Robert. *History of Propellers and Steam Navigation. With Biographical Sketches of the Early Inventors.* New York, 1851.

[McLane Report]. *Documents Relative to the Manufactures in the United States* . . . . House Executive Documents 222, 223, 22d Congress, 1st session. 2 volumes, Washington, D.C., 1833.

M'Murtrie, H[enrico]. *Sketches of Louisville And Its Environs* . . . . Louisville, 1819.

Manning, Charles H. "Comparative Cost of Steam and Water Power," *Transactions of the American Society of Mechanical Engineers*, volume 10 (1889), pages 499-508.

[Mapes, James J.] "Address, by Professor James J. Mapes," *Transactions of the Rhode Island Society for the Encouragement of Domestic Industry, in the Year 1853.* Providence, 1853.

Marestier, Jean Baptiste. *Memoir on Steamboats of the United States of America*, translated by Sidney Withington. Mystic, 1957.

Martin, Edwin T. *Thomas Jefferson: Scientist.* New York, 1961.

Martin, Margaret Elizabeth. *Merchants and Trade of the Connecticut River Valley, 1750-1820.* Northampton, 1939.

Mease, James. *The Picture of Philadelphia, Giving an Account of its Origin, Increase and Improvements in Arts, Sciences, Manufactures* . . . . Philadelphia, 1811.

Melish, John. *Travels through the United States of America, in the Years 1806 and 1807, and 1809, 1810, and 1811* . . . . Belfast, 1818.

"Memoir of Thomas Gilpin," *Pennsylvania Magazine of History and Biography*, volume 49 (October 1925), pages 289-328.

Memorial of Matthias W. Baldwin. Philadelphia, 1867.

Mitchell, Broadus. *The Rise of Cotton Mills in the South.* Baltimore, 1921.

[Mitchell, Samuel Augustus]. *Illinois in 1837* . . . . Philadelphia, 1837.

Morey, Samuel. "An account of a new explosive engine, generating a power that may be substituted for that of the steam engine," *American Journal of Science,* volume 11, no. 1 (1826), pages 104-110.

Morrison, John H. *History of American Steam Navigation.* New York, 1958.

Mumford, Lewis. *Art and Technics.* New York, 1960.

Musson, A. E., and E. Robinson. "The Early Growth of Steam Power," *Economic History Review,* volume 11, 2d series (April 1959), pages 418-439.

———. "The Origins of Engineering in Lancashire," *Journal of Economic History,* volume 20 (June 1960), pages 209-233.

"Nathan Reed," *Historical Collections of the Essex Institution,* volume 1 (November 1859), page 184.

Nevins, Allan. *Abram S. Hewitt: With Some Account of Peter Cooper.* New York, 1935.

North, Douglass C. *The Economic Growth of the United States, 1790-1860.* Englewood Cliffs, 1961.

Nowland, John H. B. *Early Reminiscences of Indianapolis, with Short Biographical Sketches of Its Early Citizens, and of a Few of the Prominent Business Men of the Present Day.* Indianapolis, 1870.

[Ogden, Aaron]. *Autobiography of Colonel Aaron Ogden, of Elizabethtown.* Paterson, 1893.

"The Oldest Steam Engine in the United States," *Engineering News,* volume 30 (9 November 1893), page 370.

Oliver, John W. *History of American Technology.* New York, 1956.

Olmsted, Frederick Law. *A Journey in the Seaboard Slave States, with Remarks on Their Economy.* New York, 1859.

Overman, Frederick. *The Manufacture of Iron, In All its Various Branches.* 2d edition, Philadelphia, 1851.

Patten, Joseph H. "Notice of the Steam-Boat Babcock," *American Journal of Science,* volume 12, no. 1 (1827), pages 115-118.

Pearse, John B. *A Concise History of the Iron Manufacture of the American Colonies up to the Revolution, and of Pennsylvania Until the Present Time.* Philadelphia, 1876.

Peck, F. Taylor. A History of the Naval Gun Factory, Washington, D.C. (unpublished doctoral dissertation, Georgetown University, 1950).

Peck, J. M. *A Gazetteer of Illinois* . . . . Philadelphia, 1837.

Perry, Amos. *Memorial of Zachariah Allen, 1795-1882.* Cambridge, 1883.

Porter, Charles T. "Reminiscences of Early High-Speed Steam Engineering," *Engineering News,* volume 44 (6 December 1900), pages 394-396.

Potter, Alonzo. *The Principles of Science Applied to the Domestic and Mechanic Arts, and to Manufactures and Agriculture* . . . . Boston, 1841.

Prager, Frank D. "An Early Steamboat Plan of John Fitch," *Pennsylvania Magazine of History and Biography,* volume 79 (January 1955), pages 63-80.

Preble, George Henry. *A Chronological History of the Origin and Development of Steam Navigation, 1543-1882.* Philadelphia, 1883.

Prime, Alfred Coxe, compiler. *The Arts and Crafts in Philadelphia,*

Maryland, and South Carolina, 1721-1785. No place, 1929.

________. *The Arts and Crafts in Philadelphia, Maryland, and South Carolina, 1786-1800: Series Two.* No place, 1932.

Proposals for Establishing a Company, For the purpose of employing the surplus power of the STEAM-ENGINE Erected near the river Schuylkill: Under the Title and Firm of the PHILADELPHIA ROLLING COMPANY. No place, [circa 1800].

Proudfit, Margaret Burden. *Henry Burden: His Life and A History of His Inventions Compiled from the Public Press.* Troy, 1904.

Quinby, A. B. "On Crank Motion," *American Journal of Science,* volume 7, no. 2 (1824), pages 316-323.

________. "On Crank Motion, in reply to the remarks of the author of a Review in the North American," *American Journal of Science,* volume 9, no. 2 (1825), pages 317-324.

Rae, John B. "The 'Know-How' Tradition: Technology in American History," *Technology and Culture,* volume 1 (Spring 1960), pages 139-150.

Read, David. *Nathan Read: His Invention of the Multi-Tubular Boiler and Portable High-Pressure Engine, and Discovery of the True Mode of Applying Steam-Power to Navigation and Railways.* New York, 1870.

[Read, Nathan]. "Autobiography of Hon. Nathan Read," *New-England Historical and Geneological Register,* volume 50 (October 1896), pages 434-436.

Redlich, Fritz. "The Business Activities of Eric Bollmann: An International Business Promoter, 1797-1821," *Bulletin of the Business Historical Society,* volume 17 (November 1943), pages 81-91; (December 1943), pages 103-112.

________. *Essays in American Economic History: Eric Bollman and Studies in Banking.* New York, 1944.

________. "The Leaders of the German Steam-Engine Industry During the First Hundred Years," *Journal of Economic History,* volume 4 (November 1944), pages 121-148.

________. "The Philadelphia Water Works in Relation to the Industrial Revolution in the United States," *Pennsylvania Magazine of History and Biography,* volume 69 (July 1945), pages 243-256.

________. "William Jones and His Unsuccessful Steamboat Venture of 1819," *Bulletin of the Business Historical Society,* volume 21 (November 1947), pages 125-136.

Reiser, Catherine Elizabeth. *Pittsburgh's Commercial Development, 1800-1850.* Harrisburg, 1951.

"Report from the Select Committee of the House of Commons, on Steam Navigation to India; with the Minutes of Evidence, Appendix, and Index. 1834," *Edinburgh Review,* volume 60 (January 1835), pages 445-482.

Report of the Commissioner of Patents, For the Year 1843, House Document 177, 28th Congress, 1st session.

Reports and Documents Upon the Subject of the Explosions of Steamboat Boilers; presented to the House of Representatives, May 18, 1832, House Document 478, 22d Congress, 2d session. Washington, D.C., 1833.

"Resolution of Mr. Beekman on the Petition of Levinus Vanderheyden, Neziah Bliss and Joseph D. Monnell," *Documents of the Senate of the State of New-York, Seventy-Sixth Session. 1853,* Document 68, volume 2. Albany, 1853.

Ritson, J. A. S. "Metal and Coal Mining, 1750-1875," *A History of*

Technology, edited by Charles Singer et al., volume 4 (New York, 1958), pages 64-98.

[Robson, Charles]. *The Biographical Encyclopaedia of New Jersey of the Nineteenth Century.* Philadelphia, 1877.

———. *The Manufactories and Manufacturers of Pennsylvania of the Nineteenth Century.* Philadelphia, 1875.

Roll, Erich. *Early Experiment in Industrial Organization: Being a History of the Firm of Boulton and Watt, 1775-1805.* London, 1930.

Rolt, L. T. C. *Thomas Newcomen: The Prehistory of the Steam Engine.* Dawlish and London, 1963.

Rowe, Frank H. *History of the Iron and Steel Industry in Scioto County, Ohio.* Ohio Historical Collections, volume 10, Columbus, 1938.

[Royall, Anne]. *Sketches of History, Life, and Manners, in the United States.* New-Haven, 1826.

Rumsey, James. *A Short Treatise on the Application of Steam* .... Philadelphia, 1788.

Rush, Benjamin. *Letters of Benjamin Rush,* edited by L. H. Butterfield. 2 volumes, Philadelphia, 1951.

Scharf, J. Thomas, and Thompson Westcott. *History of Philadelphia, 1609-1884.* 3 volumes, Philadelphia, 1884.

Schoepf, Johann David. *Travels in the Confederation [1783-1784],* edited by Alfred J. Morrison. 2 volumes, Philadelphia, 1911.

Sellers, Coleman. "President's Address, 1886," *Transactions of the American Society of Mechanical Engineers,* volume 8 (1887), pages 676-696.

Shattuck, Lemuel. *Report of the Sanitary Commission of Massachusetts, 1850.* Cambridge, 1948.

Shaw, Ralph R. *Engineering Books Available in America Prior to 1830.* New York, 1933.

[Silliman, Benjamin]. *Manual on the Cultivation of Sugar Cane, and the Fabrication and Refinement of Sugar.* Washington, D.C., 1833.

Sinclair, Bruce. *Early Research at the Franklin Institute: The Investigation into the Causes of Steam Boiler Explosions, 1830-1837.* Philadelphia, 1966.

Sitterson, J. Carlyle. *Sugar Country: The Cane Sugar Industry in the South, 1753-1950.* Lexington, 1953.

Smith, Ophia D. "Cincinnati: From Keelboat to Steamboat," *Bulletin of the Historical and Philosophical Society of Ohio,* volume 15 (October 1957), pages 259-289.

"Some Letters of Robert Fulton Concerning the Engines for His First Steamboat," *Engineering News,* volume 39 (21 April 1898), page 255.

"Southern and Western Manufactures. Relative Cost of Steam and Water Power for Manufacturing Purposes," [DeBow's] *Commercial Review,* volume 7 (August 1849), pages 128-134.

Spence, Clark C. *God Speed the Plow: The Coming of Steam Cultivation to Great Britain.* Urbana, 1960.

Stevenson, David. *Sketch of the Civil Engineering of North America.* 2d edition, London, 1859.

Stewart, James. *Steam Engineering on Sugar Plantations, Steamships, and Locomotive Engines.* New York, 1867.

Stuart, Charles B. *Lives and Works of Civil and Military Engineers of America.* New York, 1871.

Subject-Matter Index of Patents for Inventions issued by the United States Patent Office from 1790 to 1873, Inclusive. 3 volumes, Washington, 1874.

Sullivan, J. L. "Mr. Sullivan on the Revolving Engine; in reply to Mr. Doolittle," *American Journal of Science,* volume 2 (April 1820), pages 106-114.

_______. "Notice of the Revolving Steam Engine," *American Journal of Science,* volume 5, no. 1 (1822), pages 144-146.

_______. "On the Revolving Steam-Engine, recently invented by Samuel Morey, and Patented by him on the 14th July, 1815, with four Engravings," *American Journal of Science,* volume 1 (1818), pages 157-168.

_______. *Explanation, by John L. Sullivan, of the Nature of Certain Grants to Him for the Use of Steam Boats on Connecticut River* . . . (no place, 1818), p. 31.

Swank, James M. *History of the Manufacture of Iron in All Ages, and Particularly in the United States from Colonial Times to 1891* . . . . Philadelphia, 1892.

Tanner, H. S. *A Description of the Canals and Rail Roads of the United States, comprehending Notices of All the Works of Internal Improvement throughout the Several States.* New York, 1840.

Taylor, George Rogers. *The Transportation Revolution, 1815-1860.* New York, 1951.

Temin, Peter. *Iron and Steel in Nineteenth-Century America: An Economic Inquiry.* Cambridge, 1964.

_______. "Steam and Waterpower in the Early Nineteenth Century," *Journal of Economic History,* volume 26 (June 1966), pages 187-205.

Thompson, Holland. *From the Cotton Field to the Cotton Mill: A Study of the Industrial Transition in North Carolina.* New York, 1906.

Thurston, George H. *Pittsburgh As It Is; or, Facts and Figures, Exhibiting the Past and Present of Pittsburgh, Its Advantages, Resources, Manufactures, and Commerce.* Pittsburgh, 1857.

Thurston, Robert H. *A History of the Growth of the Steam-Engine.* Centennial edition, Ithaca, 1939.

_______. "Steam and Its Rivals," *Forum,* volume 5 (May 1888), pages 341-352.

Trollope, Frances. *Domestic Manners of the Americans,* edited by Donald Smalley. New York, 1949.

Turnbull, Archibald Douglas. *John Stevens: An American Record.* New York, 1928.

Turner, Ella May. *James Rumsey: Pioneer in Steam Navigation.* Scottdale, Pennsylvania, 1930.

Tyler, David B. *The American Clyde: A History of Iron and Steel Shipbuilding on the Delaware from 1840 to World War I.* Newark, Delaware, 1958.

_______. *Steam Conquers the Atlantic.* New York, 1939.

U.S. Bureau of the Census. *Historical Statistics of the United States, 1789-1945.* Washington, D.C., 1949.

Wade, John Donald. *Augustus Baldwin Longstreet: A Study of the Development of Culture in the South.* New York, 1924.

Wade, Richard C. *The Urban Frontier: The Rise of Western Cities, 1790-1830.* Cambridge, 1959.

Wansey, Henry. *An Excursion to the United States of North America, in the Summer of 1794.* London, 1798.

[Ward, James H.] *Steam for the Million. An Elementary Outline Treatise on the Nature and Management of Steam, and the Principles and Arrangement of the Engine.* Philadelphia, 1847.

Ward, Minus. *Remarks, Propositions and Calculations, Relative to A Rail-*

Road and Locomotive Engines to be used upon the same, from Baltimore to the Ohio River. Baltimore, April 1827.

________. "Ward's Alternating Steam-Engine. Some account of a Steam-Engine, called the Alternating Steam-Engine, invented by Minus Ward, of South-Carolina. In a Letter from the Inventor to the Editor," *American Journal of Science,* volume 4, no. 1 (1822), pages 90-102.

Ware, Caroline F. *The Early New England Cotton Manufacture: A Study in Industrial Beginnings.* Boston, 1931.

[Washington, George]. *The Diaries of George Washington, 1748-1799,* edited by John C. Fitzpatrick. 4 volumes, Boston, 1925.

[Washington, George]. *The Writings of George Washington from the Original Manuscript Sources,* edited by John C. Fitzpatrick. 39 volumes, Washington, D.C., 1931-1944.

Waters, Asa H. *Biographical Sketch of Thomas Blanchard and His Inventions.* Worcester, 1878.

Watmough, James H., editor "Letters of James H. Watmough to His Wife," *Pennsylvania Magazine of History and Biography,* volume 29 (1905), pages 31-43, 180-190, 296-309.

Watson, John F. *Annals of Philadelphia, and Pennsylvania, in the Olden Time . . . .* 3 volumes, Philadelphia, 1899.

Webber, Samuel. *Manual of Power for Machines, Shafts, and Belts. With the History of Cotton Manufacture in the United States.* Revised edition, New York, 1879.

Westcott, Thompson. *The Life of John Fitch, the Inventor of the Steamboat.* Philadelphia, 1857.

________. "Steam Navigation, in 1798. Rooseveldt: Livingston: Fulton," *Historical Magazine,* volume 1, 3d series (January 1872), pages 23-24.

White, John H. *Cincinnati Locomotive Builders, 1845-1868.* United States National Museum Bulletin 245. Washington, D.C., 1965.

Whitworth [Joseph], and [George] Wallis. *The Industry of the United States in Machinery, Manufactures, and Useful and Ornamental Arts.* London, 1854.

"Who First Introduced the Use of Expansive Steam for Marine Engines?" *American Repertory,* volume 4 (October 1841), pages 171-174.

Wik, Reynold M. *Steam Power on the American Farm.* Philadelphia, 1953.

Wilkeson, Samuel, Jr. "Biographical Sketch," *Publications of the Buffalo Historical Society,* volume 5 (1902), pages 135-145.

[Wilkinson, David]. "David Wilkinson's Reminiscences," *Transactions of the Rhode Island Society for the Encouragement of Domestic Industry. In the Year 1861* (Providence, 1862), pages 100-111.

________. "Reminiscences," *Eureka,* volume 1 (February 1847), pages 86-88.

Willson, Seelye A. "The Growth of Pittsburgh Iron and Steel," *Magazine of Western History,* volume 2 (October 1885), pages 540-571.

Wilson, J. Walter. "Joseph Brown, Scientist and Architect," *Rhode Island History,* volume 4 (July 1945), pages 67-79; (October 1945), pages 121-128.

[Woodbury Report]. *Letter from the Secretary of the Treasury, Transmitting . . . information in relation to Steam-Engines, &c.* House Document 21, 25th Congress, 3d session. Washington, D.C., 1838.

Wyman, Morrill. "Memoir of Daniel Treadwell," *Memoirs of the American Academy of Arts and Sciences,* volume 11 (1888), pages 325-524.

Index